Astronomy and Planetary Science - From Cryovolcanism to Black Holes and Galactic Evolution

Edited by Yann-Henri Chemin

Published in London, United Kingdom

Astronomy and Planetary Science - From Cryovolcanism to Black Holes and Galactic Evolution
http://dx.doi.org/10.5772/intechopen.101290
Edited by Yann-Henri Chemin

Contributors
Peter Y.P. Chen, Xiaolong Yang, Katherine Villavicencio Valero, Emilio Ramírez Juidías, Aina Àvila Bosch, Georg Hildenbrand, Michael Hausmann, Klaus Paschek, Myriam Schäfer, Costecia Ifeoma Onah, Augustine A. Ubachukwu, Finbarr C. Odo, Yann-Henri H. Chemin

First published in London, United Kingdom, 2022 by IntechOpen
IntechOpen is the global imprint of INTECHOPEN LIMITED, registered in England and Wales, registration number: 11086078, 5 Princes Gate Court, London, SW7 2QJ, United Kingdom

British Library Cataloguing-in-Publication Data
A catalogue record for this book is available from the British Library

Additional hard and PDF copies can be obtained from orders@intechopen.com

Astronomy and Planetary Science - From Cryovolcanism to Black Holes and Galactic Evolution
Edited by Yann-Henri Chemin
p. cm.
Print ISBN 978-1-80356-119-6
Online ISBN 978-1-80356-120-2
eBook (PDF) ISBN 978-1-80356-121-9

Meet the editor

Dr. Yann-Henri Chemin has a Ph.D. in Remote Sensing Applications (AIT, Th, 2006). He graduated in Planetary Sciences with Astronomy in 2016, after a full academic path in Earth Observation in Asia. He now concentrates on security, defense, and the space industry, a field filled with both commercial endeavors and highly technical advances. He previously worked on the hyperspectral analysis of the Apollo 12 landing site and its vicinity on the moon's surface. He also worked on the thermodynamic energy balance of the surface-atmosphere interactions of the moon Titan and on the mapping of craters on the dwarf planet called Ceres. Dr. Chemin's interests include challenges to settling on the moon and Mars, especially food creation, settlement construction, and in situ resource utilization.

Contents

Preface

In the wake of the activation of the James Webb Space Telescope, its large step in spatial resolution, and its varied radiometric capacities, the field of astronomy has witnessed breakthroughs in the frontier of analysis from even the earliest exploitation of its data. Already, the direct viewing of an exoplanet and the detection of CO_2 in the atmosphere of an exoplanet are prime examples of furthering the exploration capacity of astronomy. A new era of understanding our universe is now upon us!

Astronomy is not only about looking deep into the limit of our technological capacities to observe the farthest in the Universe, its stars and galactic evolutions, black holes, quasars, pulsars or cosmic microwave background, to name a few. The Sun orbiter, for example, is a space probe that, this year only, permitted us to return the highest resolution information from our own star yet. New evidence of heliomagnetism processes appears simply by such improved viewing capacity. In a similar way, space probes sent into our solar system, are allowed to visit a large number of planetary bodies, rocky or gaseous, alongside their moons and icy rings. Many dwarf planets, asteroids and comets have now been visited by space probes, unraveling further parts of the galactic history.

The first section of this book discusses the universe. Chapter 1 is the Introductory Chapter. Chapter 2 by Onah et al. models the temporal evolution of extra-galactic radio sources in a quasar/galaxy unification scheme. Chapter 3 by Yang discusses unified models of black hole accretion. Chapter 4 by Chen discusses the wave propagation theory and its incoherence with the Big Bang theory.

The second section of the book examines the solar system, with a focus on planetary science. Chapter 5 by Dr. Georg Hildenbrand et al. reviews cryovolcanism in the solar system. Chapter 6 by Fillavicencio et al. uses Enceladus jet plumes composition to investigate the similarity of early oceans on Earth to study the support of early forms of life.

Yann-Henri Chemin
Joint Research Centre,
European Commission,
Ispra, Italy

Section 1

Astronomy

Chapter 1

Introduction Chapter: Astronomy

Yann-Henri Chemin

1. Introduction

The observation of the Sun, the solar system, and the vast regions laying outside are commonly called *astronomy*.

In the whole range of observations of our neighborhood (**Figure 1**), we study the Sun, the Moon, the rocky planets, and the gas giants. In the last decades, the space probes sent around the solar system have enhanced our exploration capacity, that is, to go from hazy photographs to high-resolution mapping of most of our planetary bodies, even of their moons. Additionally, dwarf planets (Ceres and Vesta) and asteroids have been visited and mapped. This, in itself, it is a unique civilizational achievement in terms of exploration.

In between Mars and Jupiter, the asteroid belt is found. Further and mostly after Neptune's orbit is the Kuiper belt. Eventually, the very possible Oort cloud, a reservoir of visiting comets, is vastly beyond the orbit of (90377) Sedna (**Figure 1**).

Our solar system is located in between two arms of a spiral galaxy, within what is often called a "finger" named the *Orion spur* (**Figure 2**). The center of our galaxy, the *Milky Way*, is the seat of a supermassive black hole (SMBH) called *Sagittarius A** [3].

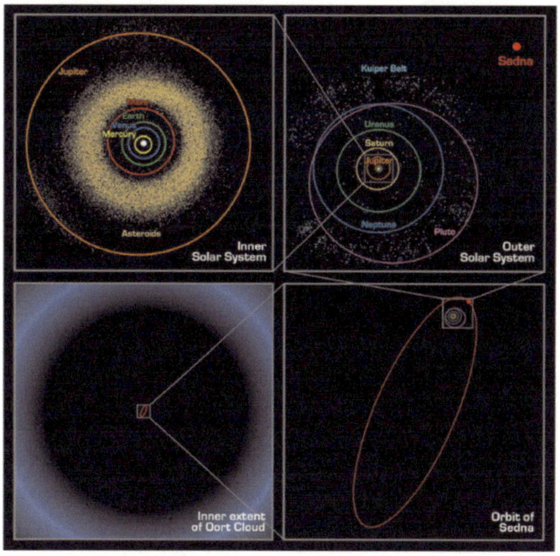

Figure 1.
Extracted from ref. [1], this is the solar system. The upper left is defined by Jupiter's orbit, upper right reaches Pluto's orbit, the Kuiper belt, and the perigee of (90377) Sedna. Bottom right is the orbit of (90377) Sedna, which is barely seen in the bottom left alongside the inner Oort cloud, which is thought to be the source of comets. Image credit: NASA/CalTech.

Figure 2.
The Milky Way. Extracted from ref. [2], image credit: NASA/Adler/U. Chicago/Wesleyan/JPL-Caltech.

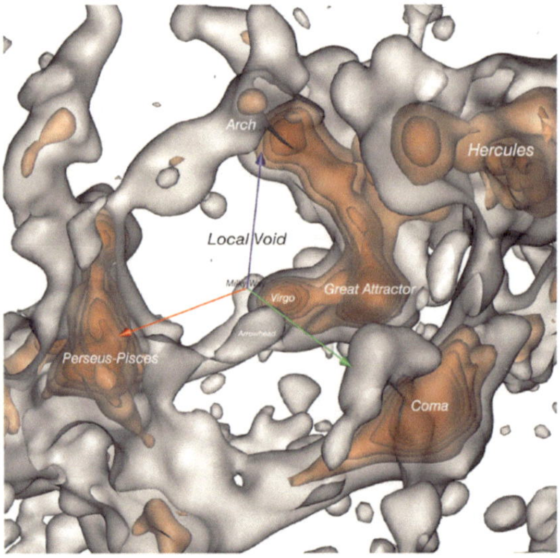

Figure 3.
Extracted from ref. [4], the local void, visualization centered on our galaxy, the milky way.

Our galaxy is located on a fringe of what could be called *mycelium* filaments. In other words, detectable matter at the cosmological level, tends to agglomerate in threads, interconnected by groups of larger material aggregations, not unlike the spread of fungal mycelium in Earth's soil, punctuated by the presence of "nodes," from which mycelium filaments extend. Of particular importance to our galaxy "suburb" is a large area void of matter (**Figure 3**). The *Local Void* has been mapped synthetically with a great resolution recently [4].

The overall observable universe is a sphere, centered on the location of the one observing. For us, it is planet Earth. When mapped, this sphere, so far, extends outwards in a radius of 46.5 billion light-years (440 Ym), as a comoving distance. The

Figure 4.
Extracted from ref. [6], scanning from the dark energy spectroscopic instrument (DESI), from the center, earth, toward the further seen so far, about one billion years after the big bang.

sloan digital sky survey (SDSS), mapping of the observable universe [5], was done not by direct distance, but in signed velocity from the observation point, deducting mostly from the red-shifting of the observations. The faster the positive red-shifting velocity, the further away the colocation to the observer (**Figure 4**).

2. Observing the sun and the stars

The closest of the stars, the Sun, is a G-IV, main-sequence star, and is located in the center of the Hertzsprung-Russell (HR) diagram [7] classification (**Figure 5**). Its next evolutions would be to leave the main sequence, that is the oblique line crossing **Figure 5** toward the upper right. It is then going to inflate and shift to become eventually a red giant, moving further up in the upper right arm of the HR diagram. Once reach maximum inflation, a cascade of gravitational collapses will happen, ejecting material by major explosions. Gravity will compact the remaining into a white dwarf, a neutron star, in the midst of its ejecta, witnessed as a *(super) nova* remnant, a nebula. The transition from the super-giant, the (super) nova reducing into the remaining white dwarf, will take the star rapidly across the HR diagram, from the upper right to the mid-upper left, and crossing down in the visible arc in the bottom left side of **Figure 5**.

The central part of the Sun is composed of a core (a fourth of its radius) where thermonuclear reactions generate energy. It has an average density ten times that of lead, and a temperature of 15 10 K. The radiative zone of the Sun is about one-third of the Sun's radius. Both transfer energy by radiative forces of photons. The pathways undergo a random walk and act as an apparent solid body [8].

The photosphere is 300–500 Km deep, and the light is emitted from there before the plasma becomes opaque. This is also the layer that defines the effective temperature of the Sun, about 5.8 10 K, plasma convection is visible there under the form of granules of sizes measured in Mm (10 m).

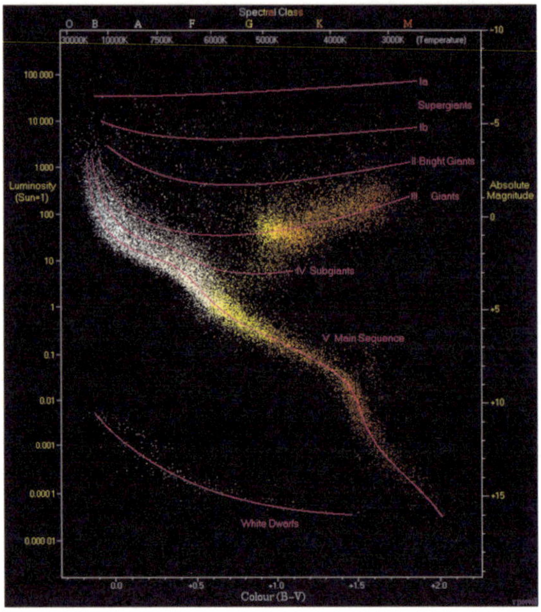

Figure 5.
HR diagram, classification of stars evolution. Image credits: Wikimedia.

Figure 6.
High-resolution image of the sun from solar orbiter, showing magnetically bound plasma. Credit: ESA & NASA/ solar orbiter/EUI team; data processing: E. Kraaikamp (ROB). [9].

The Sun's "atmosphere" (starting from **Figure 6**) is composed of the chromosphere and the corona (in that order). The chromosphere is 2000 Km deep and is the Sun's eclipse "red ring of fire." It has a steep drop in material density, and an initial temperature drop from 5.8 10^3 K to 3.5 10^3 K to eventually reach 35 10^3 K.

The corona is a very large volume above the chromosphere, vastly warmer too, made of ionized plasma of about 1106 K, with a majority of emission coming from Fe-XIV and Fe-X. It is the origin of the solar winds. Some areas with open magnetic fields yield faster solar winds (about 0.7 10^6 m/s).

We remotely sense the Sun (Solar Orbiter imagery in **Figure 6**), by analyzing electromagnetic spectra emitted from its activity. Thermodynamics and hydrodynamics applied to plasma with magnetic fields are all needed to study the radiative, convective, and exo-atmospheric conditions of the Sun energy transport.

More generally, stellar objects of different characteristics have long been observed and many physical theories have been developed relating observations and life cycles, that is, HR diagram in **Figure 5** and equations of stellar structure, respectively. Stellar oscillations [10], spherical harmonics, and resonance patterns analysis belong to geophysics and are now in common use to study and classify stars.

3. Observing the galaxies and the universe

In a similar way to stars, galaxies are also categorized along their paths of evolution. Hubble classification of galaxies evolution, the *Hubble Sequence* (**Figure 7**), reviewed here [11] (initial article [12]), provides an observation-based classification of galaxies.

The *Elliptical* galaxies start at spherical (E_0; e = 0) to the most common type of elliptical galaxies (E_7; e = 0.7). As the galaxy tends to age, central spin tends to send matter away in form of spiral arms.

The second level of classification in the *Hubble Sequence* (and further modifications) is dedicated to the extension and the shape of the *Spiral* arms of the galaxies. As seen in **Figure 7**, two branches of evolution differ in shapes of both the central bulge (whether it keeps spheroid or is barred) and the type of arms evolution. The first type, with spheroid central bulge is classified as Sa, Sb, and Sc along the evolution path. Similarly, *Barred* spiral galaxies are SBa, SBb, and SBc.

Sa (SBa) central bulge is bright and prominent.

Arms are tightly wound and smooth.

Sb (SBb) central bulge is less bright.

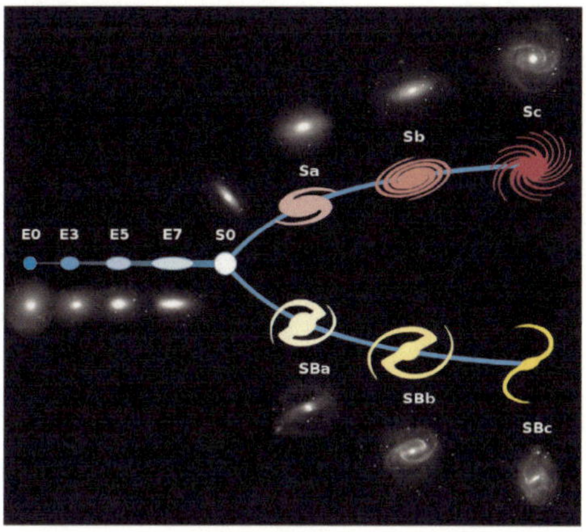

Figure 7.
Hubble classification of galaxies evolution, the Hubble sequence. Courtesy: Wikipedia.

Arms are less tight than above.

Sc (SBc) central bulge is smaller and fainter.

Arms are loosely wound (stellar clusters and nebulae).

Sd (SBd) central bulge is dim.

Arms are bright and very loose, possible fragmentary arms.

The central bulge of our galaxy, the *Milky Way*, is dominated by a super massive black hole (SMBH), Sagittarius A* [13]. Sgr A*'s event horizon image is seen in **Figure 8**. It has an estimated mass of 4.152 10^6 $M_{\odot}$ and is the prime of several stars, their orbits helping define its mass. Its observed diameter is 51.8 10^6 Km, slightly more than the Sun-Mercury maximum distance ($\odot - ☿$ = 46 10^6 Km at $\odot$ perihelion), which is about 1/3 AU (the mean distance $\odot - \oplus$).

Looking outside of the solar system has been largely enhanced with space telescopes. Furthering the capacity of the Hubble space telescope, in 2022, the James Webb Space Telescope (JWST) was activated at Lagrange 2, including the near-infrared spectrograph (NIRSpec) [14]. Its first images have been no less than revolutionary, giving direct observations of exoplanets and their atmosphere, but also looking further into the past of the universe.

The decades ahead of us promise the enhancement of our understanding of Sun, planets, the stars, black holes, and all other astronomical objects in our universe available to be observed. The observable universe itself just got smaller with JWST activated, and our understanding of the universe and its temporal unraveling is also furthering with every new data gathered. Time itself may also be better understood eventually, who knows?

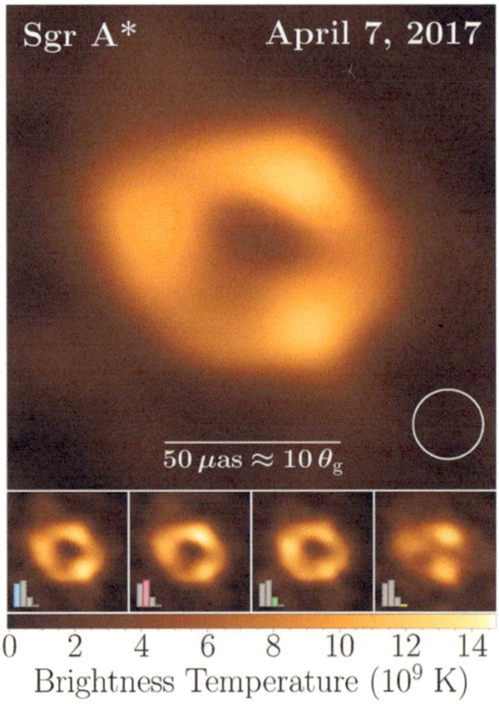

Figure 8.
The event horizon of the SMBH Sgr a at the center of the milky way [13].*

Abbreviations

⊙	Sun
☿	planet Mercury
⊕	planet Earth
AU	astronomical unit (150 10^6 Km)
DESI	dark energy spectroscopic instrument
HR	Hertzsprung-Russell
JWST	James Webb Space Telescope
NASA	US National Aeronautical and Space Administration
SDSS	sloan digital sky survey
SMBH	super massive black hole

Author details

Yann-Henri Chemin
European Commission, Joint Research Center, Ispra, Italy

*Address all correspondence to: dr.yann.chemin@gmail.com

References

[1] Chandler CO. Chasing Tails: Active Asteroid, Centaur, and Quasi-Hilda Discovery with Astroinformatics and Citizen Science [Thesis]. Flagstaff: Northern Arizona University; 2022

[2] NASA. Milky Way and Our Location [Internet]. 2017. Available from: https://www.nasa.gov/mission_pages/sunearth/news/gallery/galaxy-location.html [Accessed: September 4, 2022]

[3] Bower GC. Focus on first Sgr a* results from the event horizon telescope. ApJL. 2022;**230**(2):L12-L21

[4] Tully RB, Pomarède D, Graziani R, Courtois HM, Hoffman Y, Shaya EJ. Cosmicflows-3: Cosmography of the local void. The Astrophysical Journal. 2019;**880**(1):24. DOI: 10.3847/1538-4357/ab2597

[5] Gott JR III, Jurić M, Schlegel D, Hoyle F, Vogeley M, Tegmark M, et al. A map of the universe. ApJ. 2005;**624**(2): 463-484. DOI: 10.1086/428890

[6] Becker A. Dark Energy Spectroscopic Instrument (DESI) Creates Largest 3D Map of the Cosmos. Vol. 510. Berkeley, US: Berkerly Lab; 2022. pp. 424-2436

[7] Hertzsprung E. On the Use of Photographic Effective Wavelengths for the Determination of Color Equivalents. Vol. 1.22(63). Postdam, DE: Publications of the Astrophysical Observatory in Potsdam; 1911

[8] Howe R. Dynamic variations at the base of the solar convection zone. Science. 2000;**287**(5462):2456-2460. DOI: 10.1126/science.287.5462.2456

[9] Phys.org. Zooming into the sun with Solar Orbiter by European Space Agency. 2022. Available from: https://phys.org/news/2022-03-sun-solar-orbiter.html [Accessed: March 25, 2022]

[10] Chaplin WJ, Kjeldsen H, Christensen-Dalsgaard J, Basu S, Miglio A, Appourchaux T, et al. Ensemble Asteroseismology of solar-type stars with the NASA Kepler Mission. Science. 2011;**332**(6026): 213-216

[11] Sandage A. The Classification of Galaxies: Early History and on-Going Developments. Annual Review of Astronomy and Astrophysics. 2005;**43**: 581-624

[12] Hubble E-P. Extragalactic nebulae. APJ. 1926;**64**:321-369

[13] Event Horizon Telescope Collaboration and 270 colleagues. First Sagittarius a* event horizon telescope results. III. Imaging of the galactic center supermassive black hole. The Astrophysical Journal. 2022;**930**:L14. DOI: 10.3847/2041-8213/ac6429

[14] Böker T, Muzerolle J, Bacinski J, Alves de Oliveira C, Birkmann S, Ferruit P, et al. In-orbit commissioning of the NIRSpec instrument on the James Webb space telescope. In: MacEwen HA, Fazio GG, Lystrup M, Batalha N, Siegler N, Tong EC, editors. Space Telescopes and Instrumentation 2016: Optical, Infrared, and Millimeter Wave. Vol. 9904. Edinburgh, UK: SPIE; 2016. p. 44. DOI: 10.1117/12.2231895

Chapter 2

Evolution of Radio Source Components and the Quasar/ Galaxy Unification Scheme

Costecia Ifeoma Onah, Augustine A. Ubachukwu and Finbarr C. Odo

Abstract

In this work, a theoretical model is developed for explanation of temporal evolution of extragalactic radio sources via beaming, orientation effects and asymmetries. Equation of the form $D \approx P^{\pm q}(1+z)^{-m}$ is used to account for the D $\sim$ P/z relation. Also, $D \approx D_0 \frac{\{(1+z)-\sqrt{1+z}\}}{(1+z)^2}$ accounted properly for $\Omega_0 = 1$ cosmology than the $\Omega_0 = 0$ counterpart in linear size versus redshift of radio sources. Similarly, $D = D_c\left[1 \mp \{\ln\left(\frac{P}{Pc}\right)\}^{1/2}\right]$ model explained redshift-luminosity relationship of extragalactic radio sources. The results from the regression analyses are q = +0.003 (r = 0.04) for sources with z < 1 and q = −1.59 (r = −0.6) for all $z \geq 1$ sources. A critical linear size, Dc of 316kpc which matches the maximum theoretical linear size, D_{max} of $0.15D_0$ at a critical redshift $z_c \sim 1$ and a critical luminosity $P_c = 26.33 WHz^{-1}$ are obtained. The indication of all these results is that the linear size of radio sources evolves up to a certain limit in D–P plane and thereafter decreases with increasing luminosity as predicted in this work.

Keywords: AGNs, ESS quasars, ESS galaxies, evolutions: temporal, cosmological and general

1. Introduction

1.1 The active galactic nucleus (AGN)

The active galactic nucleus (AGN) is the existence of energetic phenomena in the nuclei, or core regions, of galaxies that cannot be clearly and directly explained to the interactions between the stars and interstellar medium. The source of radiation of AGN is definitely believed to emanate from the gravitational potential of gas from a supermassive black hole accreting mass at the core of the host galaxy. This emitted radiation by AGN is throughout the electromagnetic spectrum. The energy radiation in AGN is non-thermal unlike spectra of stars rather it is primarily as a result of the process of synchrotron radiation. In this scenario, power-law spectra, $S_\nu \sim \nu^{-\alpha}$ and high degree of linear polarization assigned to AGN object stand as evidence of synchrotron theory. AGNs have typical luminosity in the range of 10^{33} to 10^{40}WHz^{-1} [1].

Generally, an AGN possesses peculiar properties like, intense bright and point-like nucleus, radio cores with compact flat spectrum, highly ionized gas relativistically beaming out, variable fluxes observed on a wide timescale range from minutes upwards, extremely high luminosity from the range of $10^6 - 10^{14}$ solar luminosity, narrow, broad, and sometimes without lines of emission, extended radio jets and lobes.

1.2 Classification of AGN

The naming of AGN into classes and subclasses is mostly based on the exhibition of their properties or morphologies. Some AGN classes may be similar as a result of their evolution; some may be due to observed variability of luminosities [2]. Some AGN are classified based on their viewing angle as seen by the observer, which depends mostly on the obscuration of torus [3]. This brings about an unification scheme in AGNs due to the relativistic beaming and orientation effects.

Optically, AGN may be classified as type I or type II based on the optical spectral line emission. The classification of AGN using their response to radio-loudness is preferable because of the clearer observation obtained from the radio band compared to any bands' counterpart in the electromagnetic window. In this scenario, AGN can be classified as radio-loud and radio-quiet depending on their radio brightness. When they have ratios of radio (5GHz) to optical (B-band) flux $F_5/F_B \geq 10$, they are called radio-loud, otherwise known as radio-quiet if $F_5/F_B \leq 10$ [2, 4]. Since this work is entirely based on the observed radio properties of various AGN samples plus the clarification and consideration of radio window observation, the classification of AGNs based on radio-loudness is represented in **Figure 1**.

1.3 Radio-loud AGNs

Radio-loud sources emit more energies in the radio band than in the optical waveband, and hence possess F_5/F_B fluxes ≥ 10 [2]. There are about 15–20% of

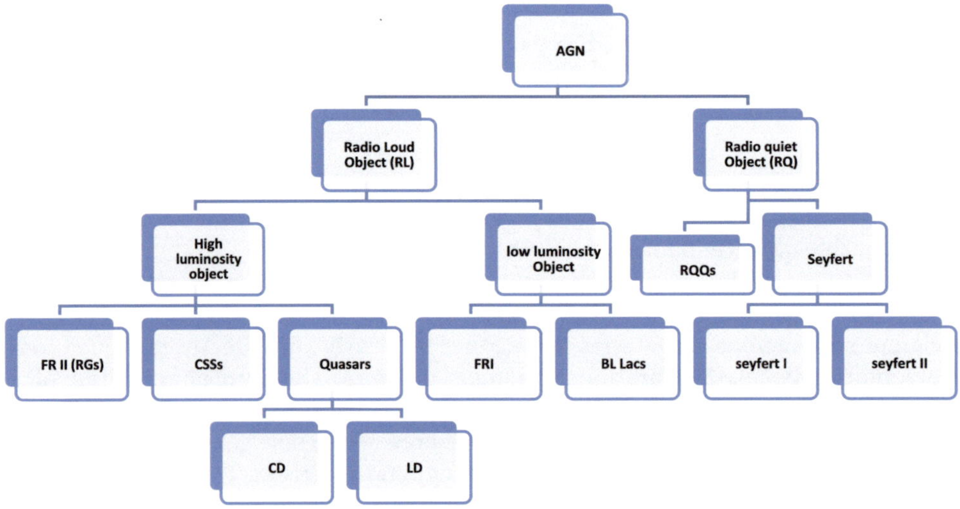

Figure 1.
Classification of AGNs based on radio-loudness.

radio-loud AGN. They are mostly elliptical galaxies in accordance with Hubble turning-fork proposition (population II)—meaning that they contain mostly of old stars with little interstellar gas. These objects are further classified into two: high luminosity objects and low luminosity objects.

1.4 High luminosity objects

They have total radio luminosity, $P_{178} \geq 10^{35} WHz^{-1}$, with a highly ionized accretion disk [5]. These comprised of radio-loud quasars, Compact Steep-Spectrum Sources (CSSs) and Fanaroff-Riley class II radio galaxies. The works are mostly on these classes of AGN.

1.4.1 Fanaroff-Riley class II radio galaxies

These are known for their high luminous intensity with the possession of extended double lobes/jets where one side is Doppler enhanced. They also have smooth jets as a result of highly supersonic flows. The jets are also edge-brightened and terminate in hotspots [5–7]. They polarize linearly with the electric field vector being perpendicularly to the jets. In a unification scheme, [8] suggested that FR II sources are misaligned counterparts of core-dominated quasars.

1.4.2 Quasars

Quasars are classified further into core-dominated and lobe-dominated quasars. These are core-dominated if the radio emissions emanate mostly from the core, otherwise it is a lobe-dominated one.

The core-dominated ones possess properties like flat radio spectra with spectra index, $\alpha \leq 0.5$, $(s_\nu - \nu^{-\alpha})$ as a result of synchrotron self-absorption mechanism, cores with extremely brightness, broad emission lines and one-sided jets/lobes. These types of quasars dominated the survey at high frequencies and high redshifts. These classes of quasars show more asymmetry than the lobe-dominated counterpart [9].

On the other hand, lobe-dominated quasars, unlike the core-dominated, have two extended lobes straddling a weak compact core. They are also high luminosity sources with total luminosity, $P_{178} \geq 10^{35} WHz^{-1}$. They are characterized by steep radio spectra $(\alpha > 0.5)$. They show spectra with broad emission lines; hence, they are referred to as broad-line region sources (BLRS). They also have higher redshifts when compared with radio galaxies [9].

1.4.3 Compact Steep-Spectrum Sources (CSSs)

The CSS sources are characterized by sharp peaks exhibited in their radio spectra. As their names imply, they are compact bright radio sources with a population up to 30% [10]. They are also called "youth" scenario [10] being believed to be the younger phases of powerful large-scale extragalactic radio sources. They have a small radio size, $D \leq 15kpc$, with a steep radio spectrum, $\alpha \geq 0.5$, a very high radio luminosity, $\log P > 10^{26} WHz^{-1}$ at frequency, $\nu = 2.7$ GHz [10]. CSS radio sources exhibit brightness temperature up to $10^{10}K$ [10]. Their radio structure is symmetric with low radio polarity and large Faraday rotation measures.

They are CSS radio galaxies if they have double lobes with weak jets and cores emitting weakly, otherwise CSS quasars if they exhibited brighter cores and jets [11]. Majority of CSS radio jets are one-sided and superluminal [11]. From the theory of orientation-based unification scheme, the morphological difference between the CSS radio galaxies and CSS quasars is that objects seen close to the line of sight of the observer are referred to CSS quasars otherwise radio galaxies [2].

1.5 Low luminosity radio sources

Low luminosity radio sources have total radio luminosity, $P_{178} < 10^{26}\,WHz^{-1}$ with less ionized accretion disks [12]. Examples of these sources include FR I radio galaxies and BL Lacertae objects.

1.5.1 Fanaroff and Riley class I (FR I) radio galaxies

According to [12], FR I radio galaxies are characterized by extended double-lobed with low frequency, $\nu \sim 400$ MHz. FR I has $R_{FR} < 0.5$ as the source faints away from the nucleus, while FR II with brightness further away from the nucleus has $R_{FR} \geq 0.5$. R_{FR} ratio is the ratio of the distance between the regions of highest surface brightness to the lowest brightness contour of the central galaxy.

Moreover, FR I sources are symmetric with smooth and continuous jets which begin as one-sided nearer the core and two-sided at a few kilo parsecs away. FR I sources are located in rich clusters that highly emit x-ray gas. The x-ray gas sweeps back and distorts the FR I radio structure as it moves across the interstellar cluster, hence giving FR I object narrow-angle-tail or wide-angle-tail according to the strength of the ram pressure of the gas [13–14].

1.5.2 BL Lacertae sources (BL Lacs)

BL Lacs objects are the most violent AGN known. They have properties like very weak or sometimes no radio emission lines, compact radio core, rapid and high peak variable fluxes, superluminal flows. They are also known as blazars just like optically violently variable (OVV) quasars.

1.6 Radio quiet objects

These are objects that emit more in the optical window than they do in the radio band. AGN sources are said to be radio quiet if they emit more of their energy in the optical waveband than in the radio waveband. They have properties like F_5/F_B fluxes ≤ 10, low luminosity at 6 cm less than $10^{26}WHz^{-1}$, short jets, few relativistic particles and total weakness of the radio sources [15]. Examples of these objects are radio quiet quasars and seyferts (seyfert I and seyfert II). In this work, it is centered more on high luminosity radio-loud objects like FR II radio galaxy and quasar.

1.7 Features observed in AGNs

The observational morphological features of a radio source are radio core, jets, lobes and hotspots, though every source may not exhibit all these features.

1.7.1 Radio core

This is the core engine where the energetic radio emission mechanism in EGRSs is assumed to originate. This core is divided into steep spectrum cores and ultra-compact flat spectrum [16]. The steep spectrum cores are characterized in some radio galaxies by having more extended radio cores of few kilo parsecs in size and steeper spectra ($\alpha \geq 0.5$), example as found in seyferts and some spiral galaxies. On the other hand, the ultra-compact cores' counterparts possess properties like 1kpc in sizes and flat radio spectra ($\alpha < 0.5$), signifying synchrotron self-absorption that arises as a result of the re-absorption of some radiate relativistic electrons within the radio source. Quasars objects' core emissions appear to be more powerful than that of the radio galaxies.

1.7.2 Jets

These are the conduits through which the high-energy particles are transporting from the cores to the other extended radio structures. A typical radio jet is expected to be at least four times as long as its width, in line with the radio core and separable from other features at high resolution [17]. The radio jets which can be one or two-sided exhibit properties like emissions with steep spectra of spectral index, $\alpha \sim 0.5$–0.9.

1.7.3 Radio lobes

These are one of the extended structures of radio sources that cover a range of hundred kilo parsec to a few mega parsecs. These are characterized by possession of non-thermal steep spectra of spectral indices, $\alpha \geq 0.5$, with a high degree of polarization at high frequency. They exhibit morphological features like tail, plumes, bridges and haloes. Tails are structures formed as a result of deflected plasma interacting with external medium, while plumes are extended regions with low luminosity that faints away from the whole source. Bridges occur in the inner lobe regions of radio galaxies, while the haloes are low surface brightness structures containing old aged plasma [18].

1.7.4 Hotspots

Hotspots are the brightest region of the extended radio structure formed at the end of the lobe where kinetic power of the jets is converted into random motion within the relativistic plasma and strengthened magnetic fields [19]. They have a linear size of 1kpc and steep spectra, $\alpha \sim 0.5$–1 slightly flatter than that of the surrounding diffuse emission [20]. See **Figure 2**.

It has been established that the appearance of EGRSs is substantially modified by relativistic beaming and orientation of the radio axes with the line of sight, leading to asymmetries in the observed radio structures. Similarly, radio sources are known to undergo some form of cosmological as well as temporal evolution. However, the amount of relativistic beaming and the nature of the evolution present in different classes and subclasses of the EGRSs are still a subject of intensive research. In particular, different source samples show a wide range of the amount and nature of temporal evolution as reported in literature. Hence, the aim of this work is to analytically examine the observed radio properties of different samples of EGRSs for radio source structural asymmetry, use relativistic beaming and source orientation model to explain any

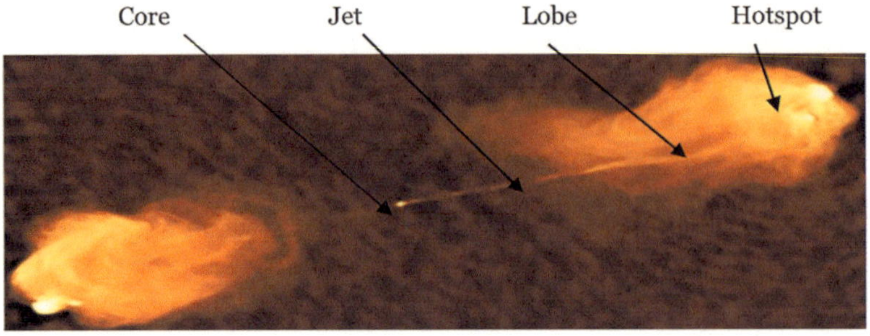

Figure 2.
Unified structure of an typical extragalactic radio source [21].

observed structural asymmetry in the radio sources and finally develop a model that can unambiguously explain the temporal evolution in extragalactic radio sources.

2. Theory of relationships

In this approach, relations between various parameters of the radio sources would be derived based on established laws and theories under certain assumptions. The theories of relationships are outlined.

2.1 Theory of temporal evolution in radio sources

The standard relationship for an ideal temporal evolution model of extragalactic radio sources can be studied using the relationship between the two key parameters known as the observed linear size (D) and the spectral luminosity (P_v) in a general power-law function as [22]:

$$D \approx D_0 P_v^{\pm q} \tag{1}$$

where D_0 (a constant) represents D at $z \sim 0$ while the slope (q) is the temporal evolution parameter.

Meanwhile, from the Friedmann-Robertson-Walker universe, the radio angular size–redshift (θ–z) relation is described [23, 24] as

$$\theta = \frac{D(1+z)^2}{d_L}, \tag{2}$$

where D can be represented using the unit of kpc and d_L is the luminosity distance expressed [24, 25] as

$$d_L = \frac{2c}{H_0 \Omega_0^2}\left\{\Omega_0 z + (\Omega_0 - 2)\left[(\Omega_0 z + 1)^{\frac{1}{2}} - 1\right]\right\}, \tag{3}$$

and c is the speed of light. Moreover, the radio luminosity (P) of a radio source at redshift (z) can be defined as a function of the spectral flux density (S_v) at observing frequency (ν) according to the [26] as:

$$P = P_0(1+z)^{\beta} \tag{4}$$

where α stands for spectral index. Thus, from [27] work, the relationship between P and z can be approximated to a simple power-law function of the form:

$$P = P_0(1+z)^{\beta} \tag{5}$$

where β is the slope of the P–z which is supposed to be constant over all values of z for a given sample of sources. In this scenario, a significant correlation for all values of z in the line with $\beta > 0$ is expected in a P–z data of different samples of extragalactic radio sources, due to luminosity selection effects in flux density limited samples [27].

Besides, researchers [23, 24, 28] suggest that linear sizes of extragalactic radio sources evolve with cosmological epoch in the form:

$$D \approx D_0(1+z)^{-k} \tag{6}$$

where D_0 is the normalized linear size which depends on the assumed cosmology and k represents the evolution parameter which could be a function of both cosmological evolution and luminosity selection effects.

Alternatively, with different values of Ω_0, the observed θ–z data of extragalactic radio sources deviate significantly from the standard Friedmann models [29], as a result of an entanglement of two effects, namely linear size evolution [29] and luminosity selection effect [30]. Hence, the linear size evolution of extragalactic radio sources can therefore be expressed as a function of both redshift and luminosity in the form of [24, 30, 31]

$$D_{(P,z)} \approx P^{\pm q}(1+z)^{-m} \tag{7}$$

where m is the residual cosmological evolution parameter defined [24, 27] as:

$$\mathrm{m} = \mathrm{k} \pm \mathrm{q}\beta \tag{8}$$

When the effect of luminosity is controlled and this depends on the product of temporal evolution (q) and luminosity selection effect (β).

It is certain from above relation that if luminosity selection effect is above cosmological evolution, then, when the effect of luminosity is corrected, $m < 0$, so that Eq. (8) becomes

$$\mathrm{k} - \mathrm{q}\beta < 0 \tag{9}$$

and $q > \frac{k}{\beta}$. For such sources, a significant positive D–P correlation is envisaged, suggesting a linear relationship between the two parameters. On the other hand, if there is a residual linear size evolution at certain values of k, $q < 0$ and $m > 0$, then $q < \frac{k}{\beta}$, indicating a significant D–P anti-correlation and also implies that the luminosity decreases with the expansion of the sources. Hence, the values of k are expected to vary in a given sample. In this scenario, the temporal evolution model for an assumed cosmology can be constrained using the D–z plane.

It is also believed that D should not increase with z for all z. Although, considering the fact that radio sources are not just rigid rods and the θ–z plane depends on the assumed cosmology, which is also, characterized by the value of Ω_0, in principle, this may not always be the case. Hence, the P–D track of all inclusive radio source samples could be defined [24, 29] as:

$$P = P_{\max} \exp\left[\pm\left(\frac{D}{D_c} - 1\right)^2\right] \tag{10}$$

where P_{max} is the maximum luminosity and D_c is the critical linear size at which the P_{max} is emitted by a radio source. Rearranging Eq. (10) gives

$$D = D_c\left[1 \mp \left\{\ln\left(\frac{P}{Pc}\right)\right\}^{1/2}\right] \tag{11}$$

Eq. (11) comprises of two separable components corresponding to $k < 0\ (-)$ and $k > 0\ (+)$. The D–P relation (c.f. Eq. (11)) is shown in **Figure 7**.

However, using Eq. (3) in (2), the linear size can be defined as:

$$D = \frac{D_0\left\{\Omega_0 z + (\Omega_0 - 2)\left[(\Omega_0 z + 1)^{1/2} - 1\right]\right\}}{(1+z)^2}, \tag{12}$$

where $D_0 = \frac{2c\theta}{H_0\Omega_0^2}$ is the intrinsic linear size and a constant. Hence, for inflationary universe, $\Omega_0 = 1$, [32], the linear size of a radio source depends on z in the form [27]:

$$D \approx \frac{D_0\{(1+z) - \sqrt{1+z}\}}{(1+z)^2} \tag{13}$$

Eq. (13) indicates two components of z: $z < 1$ and $z > 1$, where D increases with increasing z in the first so that $k > 0$, while the revised is the case in the last component for all D and k. On the other hand, assuming $\Omega_0 = 0$, on supposition of a low-density universe, for which d_L is given as [33]:

$$d_L = \frac{2c}{H_0}\left[(1+z)^2 - 1\right], \tag{14}$$

Eq. (2) yields,

$$\theta = \frac{DH_0(1+z)^2}{2c\left[(1+z)^2 - 1\right]}. \tag{15}$$

Hence, D can be expressed in relation with z as

$$D = D_0\left[1 - \frac{1}{(1+z)^2}\right] \tag{16}$$

where the intrinsic radio size, $D_0 = \frac{\theta 2c}{H_0}$. The variation of D with z for both cosmological models is shown in **Figure 3**.

It is obvious from **Figure 3** that there is increase in D as z increases up to a certain maximum point known as a critical value of $D_C \approx 0.146\ D_0$ (kpc) at a redshift

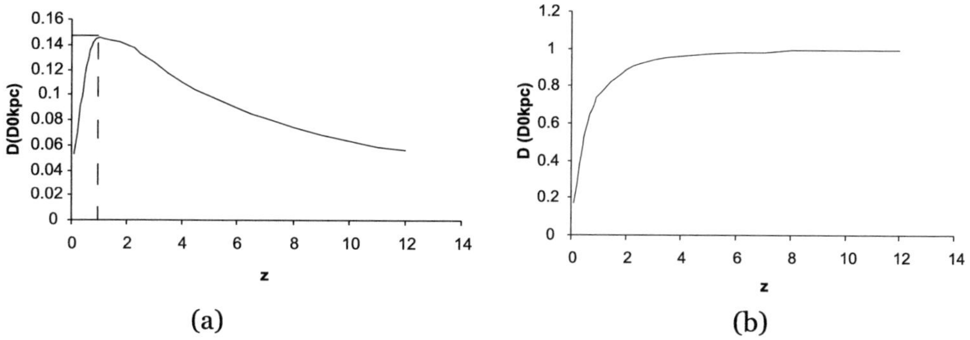

Figure 3.
Variation of D with z for $\Omega_o = 1$ (a) and $\Omega_o = 0$ (b) cosmologies.

maximum called critical $z_c \approx 1$, and decreases after for $\Omega_0 = 1$. This implies that for parameters, $z < 1$, $k > 0$ and $q > 0$, there is a positive temporal evolution, while for $z > 1$, $k < 0$ and $q < 0$ a negative temporal evolution is envisaged which is in good agreement with the predictions of Eq. (13).

On the other hand, there is increase in D up to a critical point $Dc \approx 0.8D_0$ after which it remains constant for $\Omega_0 = 0$. It is now clear that in the two cosmological models, there is indication of Dc at which luminosity is maximum, suggesting that for any assumed cosmological model, the Dc value obtained from D–P turnover can still be found using the D–z plane of the same data. Hence, in any assumed cosmological model, there will be an expected range of the Dc value of $0.15D_0$ to $0.8D_0$ bounded by the two cosmologies. In this scenario, temporal evolution in the current sample of EGRSs would be modeled in terms of the current inflationary one with $\Omega_0 = 1$.

2.1.1 Relativistic beaming based on orientation and radio source asymmetries

The standard relation for explaining an ideal relativistic beaming and orientation effects for extragalactic sources is often carried using a key parameter known as core-dominance, (R) defined [34] as:

$$R = \frac{P_C}{P_E} = \frac{P^C_{5GHz}}{P^E_{1.4GHz}} \left(1.4/5\right)^{-\alpha_E} (1+z)^{-\alpha_E} \tag{17}$$

where P_C represents core-luminosity at 5 GHz, P_E is the extended/lobe luminosity at 1.4 GHz and α_E is the lobe spectral index. However, if relativistic beaming effect at small orientation angle, then R can be expressed in terms of the jet speed (β) and inclination angle (ϕ) in the form [35]:

$$R = \frac{P_C}{P_E} = \frac{R_T}{2}\left[(1-\beta\cos\phi)^{-n+\alpha} + (1+\beta\cos\phi)^{-n+\alpha}\right] \tag{18}$$

where R_T is the value of R at $\phi = 90^\circ$ and n is a parameter that depends on the assumed flow model of the radio jet. For radiating plasma with continuous jet n = 2, otherwise n = 3 if the jet consists of blobs.

An obvious outcome of relativistic beaming and orientation effects in AGNs is the wide range of asymmetry observed in their radio structures. The radio source asymmetry can be explained using the arm-length ratio (Q), defined as the ratio of the distance, from the central engine, of a plasma element emitting radio waves on the approaching jet side (d_{app}) to that on the receding jet side (d_{rec}), [36, 37] given as:

$$Q = \frac{d_{app}}{d_{rec}} = \frac{1 + \beta \cos \phi}{1 - \beta \cos \phi} \tag{19}$$

On the other hand, [38, 39] suggested that,

$$x = \frac{Q - 1}{Q + 1} \tag{20}$$

where, x represents the index of the asymmetry.

This x parameter which believed to have better relationship with orientation when compared to Q can further be defined in terms of the viewing angle as [40, 41]:

$$x = \beta \cos \phi \tag{21}$$

The relativistic beaming in AGN at small angles to the line-of-sight is fundamentally characterized by beaming enhancement factor (δ) expressed [40, 42] as:

$$\delta = \gamma^{-1}(1 - \beta \cos \phi)^{-1} = \tag{22}$$

where, γ is the bulk Lorentz factor of the jet [8, 42, 43] defined as:

$$\gamma = \frac{1}{(1 - \beta^2)^{\frac{1}{2}}} \tag{23}$$

Also, assuming $\phi = 0^0$ in Eq. (18) and analyzing further, [44, 45] gives

$$R_{\max} \approx R_T \gamma^2 (2\gamma^2 - 1) \approx 2R_T \gamma^4 \tag{24}$$

While at $\alpha = 0$ and $\beta \sim 1$, for high luminosity radio-loud AGN sources emitting at small angle to the line-of-sight of the observer, (to a first approximation) Eq. (18) reduces to

$$\cos \phi_m = 1 - \left(\frac{2Rm}{R_T}\right)^{-\frac{1}{n}} \tag{25}$$

where R_m is the mean value of the R-distribution and ϕ_m is the mean observation angle.

Thus, it can be shown from Eq. (23) that the asymmetry parameter x can be expressed in terms of the beaming enhancement factor as:

$$x = \frac{\delta\gamma - 1}{\delta\gamma} \tag{26}$$

Eq. (26) implies that there is an association of relativistic beaming and radio source asymmetry. In asymmetric sources with $Q > 1$, x–D anti-correlation is expected if geometric projection at small viewing angles is responsible for the observed asymmetry. Following [40, 41], we assume a linear x–D relation of the form:

$$x = x_{max} - \lambda D, \tag{27}$$

where x_{max} represents the maximum x for a sample at D $\sim$ 0($\phi_c \neq 0$) to the line-of-sight [44] and λ is the slope. Thus, if $\beta \sim 1$ for the relativistic jets, analysis of Eq. (20) down to Eq. (28) for optimum beaming gives [40, 45]:

$$\phi_c \approx \sin^{-1}\left(1/_{\gamma}\right) \approx \cos^{-1} x_m, \tag{28}$$

Hence, if relativistic beaming at small viewing angles is responsible for the observed structural asymmetry, the critical viewing angle ϕ_c as well as the Lorentz factor γ can be obtained using the x–D data.

2.2 Statistical analyses and results

2.2.1 The source samples

The data used in the present analysis were drawn from a well-defined source sample of [46] compilation which contains required information on the two objects of interest–ESS quasars and ESS radio galaxies, [34] compilation of 542 extragalactic radio sources and the deep VLA sample of FSRQs compiled by [47]. In these samples, there is wide dispersion in the distributions of the observed parameters.

2.2.2 Distributions of observed radio parameters

The distributions of the linear size, D, of the extended steep-spectrum sources (ESSs) in logarithm scales are represented in **Figure 4a**. The graph shows D-values of 1896.3, 17.1 and 1879 kpc for the maximum, minimum and range respectively for ESS quasars, while D-values of 5853.30, 29.80 and 5823.5 kpc were obtained for maximum, minimum and range respectively for ESS galaxies. The entire sample yields D-values of 5853.3, 17.1 and 5836.2 kpc for maximum, minimum and range respectively. Further analyses yield median D-values of 148.90, 323.59 and 201.90 kpc for ESS quasars, ESS galaxies and entire sample respectively. The mean D-values of 144.54 $\pm$ 7.52 kpc and 288.40 $\pm$ 33.66 kpc were obtained respectively for ESS quasars and radio galaxies.

The distributions of the redshift, z of the ESSs in logarithm scales are represented in **Figure 4b**. The graph shows z-values of 2.88, 0.05 and 2.83 for the maximum, minimum and range respectively for ESS quasars, while z-values of 3.22, 0.006 and 3.21 were obtained for maximum, minimum and range respectively for ESS galaxies. The entire sample yields z-values of 3.22, 0.006 and 3.21 for maximum, minimum and range respectively. Further analyses yield median z-values of 1.89, 1.17 and 1.62 for ESS quasars, ESS galaxies and entire sample respectively. The mean z-values of 1.95 $\pm$ 0.01 and 1.30 $\pm$ 0.02 were obtained respectively for ESS quasars and radio galaxies.

The distributions of the radio luminosity, P of the ESSs in logarithm scales are represented in **Figure 4c**. The graph shows P-values of 27.90, 25.71 and 2.18 WHz^{-1} for the maximum, minimum and range respectively for ESS quasars, while P-values of 28.01, 24.97 and 3.04 WHz^{-1} were obtained for maximum, minimum and range respectively for ESS galaxies. The entire sample yields P-values of 28.01, 24.97 and

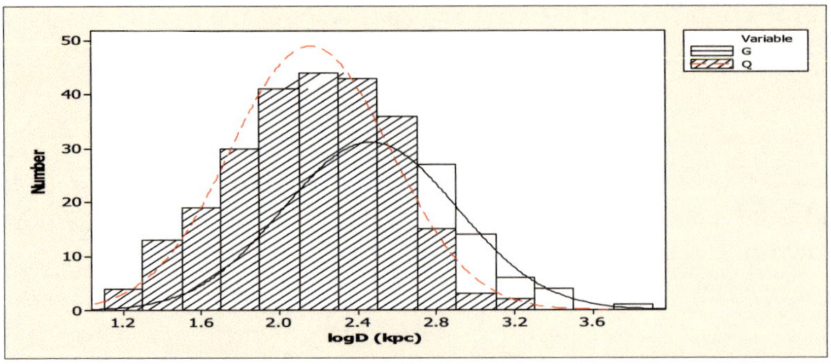

(a)

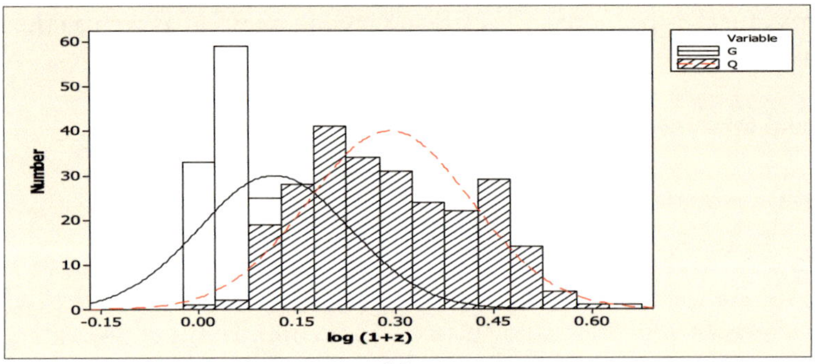

(b)

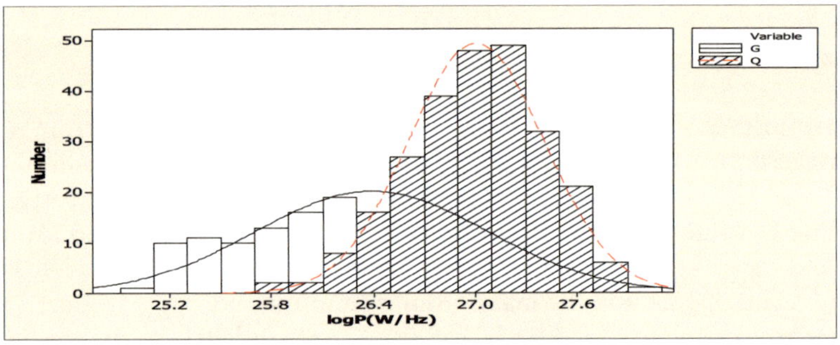

(c)

Figure 4.
Distribution of (a) D, (b) z and (c) P respectively for ESS quasars (lines) and ESS galaxies (plane).

3.04 WHz^{-1} for maximum, minimum and range respectively. Further analyses yield median P-values of 27.04, 26.39 and 26.84 WHz^{-1} respectively for ESS quasars, ESS galaxies and the entire data. The mean P-values of 27.01 ± 0.01 WHz^{-1} and 26.39 ± 0.02 WHz^{-1} were obtained respectively for ESS quasars and radio galaxies **Figure 4**.

2.2.3 D: P/z correlation

Figure 5 represents the scatter plot of linear size, D against the redshift, z. The median value data in seven redshift bins is superimposed on the plot. Critical investigation of the plot shows that on average, the linear size increases with increasing redshift up to a value logDc = 2.5 kpc (Dc = 316.23 kpc) at $z_c = 1$, after which it

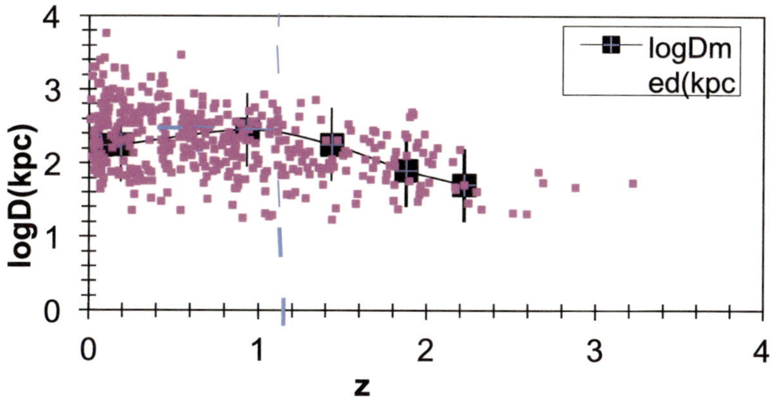

Figure 5.
Scatterplot of logD (kpc) against z for entire sources (circle) with median values (square) superimposed [24].

Parameters	D–P			D–z		
	D_0	q	r	D_0	k	r
$z<1$	2.37	0.003	0.004	2.55	−1.09	−0.12
$z\geq1$	2.67	−1.59	−0.6	−1.59	2.67	−0.5
$P<P_c$	4.91	0.17	0.21	2.51	0.81	−0.05
$P\geq P_c$	16.48	−0.32	−0.7	2.63	−1.50	−0.6

Table 1.
D–P/z regression analyses results for both $z < 1$ and $z\geq1/P<P_c$ and $P\geq P_c$ [24].

decreases with increasing redshift [24]. This is in agreement with the prediction made in **Figure 3(a)**. Hence, the present data obviously proved consistency with the inflationary model of the universe ($\Omega_0 = 1$). The median values give significant trends with correlation coefficients of +0.95 and − 0.90 for $z_c = 1$ and $z_c \geq 1$ respectively. Results of the regression analyses of the D–P/z data for $z_c = 1$ and $z_c \geq 1$ are summarized in **Table 1**.

In modelling the temporal evolution of the sample, **Figure 6** represents the scatter plot of projected linear size (D) and the radio luminosity (P). Similarly, the median value data in eight uniform luminosity bins are superimposed on the plot. There is an obvious trend indicating that the linear size first increases

with increasing luminosity up to a certain value and thereafter decreases with increasing luminosity. The median value data showed very significant trends. This suggests that the turnover occurs at a critical point of luminosity, $\log P_c =$ 26.33 WHz^{-1} and logDc = 2.51 kpc (316.23 kpc) [24]. A summary of the results of these regression analyses of the D–P and D–z data for sources with $P \leq Pc$ and $P > Pc$ is presented in **Table 1**.

The results in **Table 1** did not show any obvious trend in D–P relation for sources with $P < P_c$. However, for sources with $P \geq P_c$, there is a fairly strong correlation. Therefore, the weak trend found in the region below z = 1 and $P < P_c$ is believed to be due to the effects of luminosity selection. In this scenario, the low redshift samples, $z < 1$ have more impacts on average luminosity-redshift plane than the high redshift, $z \geq 1$ counterparts in any flux density limited samples.

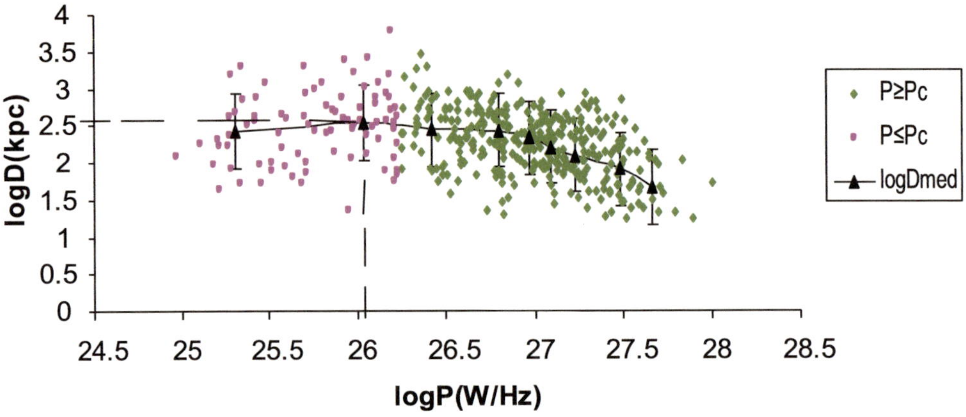

Figure 6.
Scatterplot of logD(kpc) against logP (WHz^{-1}) *for both sources with* $P \leq P_c$ *(circle),* $P \geq P_c$ *(square) and median (triangle) values superimposed [24].*

2.2.4 Luminosity selection effect on temporal evolution model

Figure 7 represents the P–z scatter plot of the sample. There is a turnover at critical point $\log P_c = 26.33$ W/Hz and $z \sim 0.05$ in the P–z plane. This point of P is presumably the value of luminosity in **Figure 6** that will correspond to D_c at $z_c \sim 1$. However, the critical luminosity, P_c found in the P–z plane is inconsistent with sources at z = 1. Hence, the luminosity-redshift relation of the current data did not assume $\Omega_0 = 1$ model but $\Omega_0 = 0$ cosmological model (low-density universe) [24].

In **Figure 6**, the effect of Eqs. (10) and (11) in the light of the temporal evolution model was considered. Hence, the data are grouped into two; P < Pc and P ≥ Pc. The results of the regression analyses are shown in **Table 1**. These opine that the linear size, D of radio sources increases up to critical radio luminosity, $\log Pc = 26.33 WHz^{-1}$ and decreases thereafter. This is in agreement with Eq. (10), hence suggesting that eqn. (10) is approximately correct to zero order. This is applicable when critical linear size Dc is used. Dc = 316kpc is obtained from the D - z plane at the turning point. This Dc corresponds to theoretical $D_c = 0.14D_0$ at $z_c = 1$ for $\Omega_0 = 1$ in **Figure 3a** and $\log Pc = 26.33\ WHz^{-1}$ in **Figure 7**. The indication of this is that D_0 ($\sim$ 2100 kpc in the

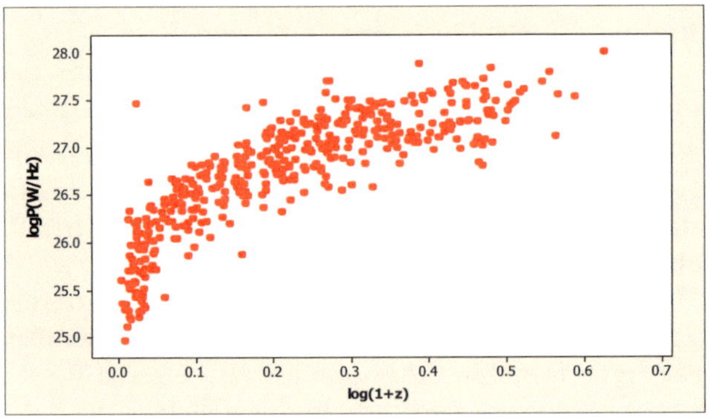

Figure 7.
Scatterplot of logP (W/Hz) against log (1 + z) for all sample.

observed data) is approximately the linear size at the earliest epoch (at z $\sim$ 0.02 in the observed data of the present sample). The correlation coefficient, r $\sim$ +0.4, -0.5 and $-$ 0.9 respectively for sources with P < Pc, P $\geq$ Pc and the median values respectively were obtained. These suggest that there are positive and negative correlations in the D–P track at P < Pc and P $\geq P_c$ respectively, indicating that the D_c = 316kpc of the observed samples is consistent with $D_c = 0.14D_0$ just in accordance with the theory for the inflationary model of the universe, $\Omega_0 = 1$ only, at $z_c = 1$ and P_c = 26.33WHz^{-1}, proving Eq. (11) to be perfectly correct. Hence, temporal evolution in extragalactic radio sources can be explained.

2.2.5 The x – *D relationship*

There is a fairly significant trend in the *x*–D plot, which is obvious at the upper envelope function. This yields: x = 0.35–0.0006 D with a correlation coefficient r $\sim -$ 0.5, chance probability $\rho \sim 10^{-10}$ and critical viewing angle, $\phi_c \approx 70^o$, which corresponds to $\gamma = 1.1$. The upper envelope function gave correlation coefficient, r $\sim$ 0.9, $\phi_c \approx 48^o$ and $\gamma = 1.3$. The analyses for separate objects, radio galaxy and quasar sub-samples, of upper envelope functions yield $\phi_c \approx 59^o$; γ = 1.2 and $\phi_c \approx 33^o$; $\gamma = 1.8$, for radio galaxies and quasars, respectively. These results correspond to angular separation $\left(\phi_{sep}\right)$ of $\sim 26^o$, based on the upper envelope functions [40]. The results are shown in **Table 2** and **Figure 8**.

The results opine that the relativistic beaming and source orientation effects are the major cause of large-scale structural asymmetries observed among powerful

Source parameters	Radio galaxies	Quasars	Galaxies + quasars
x_m	0.16	0.25	0.20
λ	+0.000056	−0.0002	+0.000002
r	+0.3	−0.3	−0.5
Sig. (%)	5.0	5.0	5.0

Table 2.
x–D regression analyses results [40].

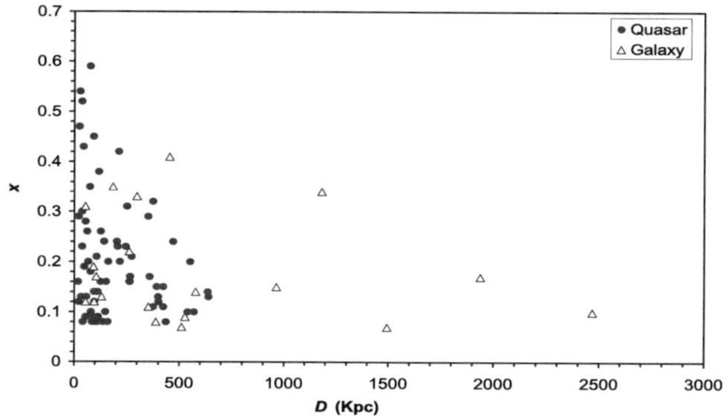

Figure 8.
Scatterplot of the fractional separation difference (x) as against projected linear size (D) for quasars (•) and radio galaxies (Δ) [40].

extragalactic radio sources, which are more obvious in core-dominated sources with large core-to-lobe luminosity ratio.

2.3 General discussions and conclusions

2.3.1 Discussions

The study of the redshift effect dependence of radio size in extragalactic radio sources has been of great interest since inception of the universe. This is obviously important in cosmological studies more especially in testing of the evolution in the extragalactic radio sources. In this scenario, a theoretical model that can best interpret the D–P track is developed. According to [48] the analyzes obtained, the high redshift radio sources show high bending angles, distorted and smaller structures than the low redshift radio sources. According to the researchers, the effects on the radio source evolution depend mostly on intracluster/ or interstellar medium. A plausible model was first developed by [49] for strong linear size evolution for radio galaxies. This model explained that the typical radio sources, both giant galaxies and sub-galactic quasars, evolve at high redshifts, z.

The temporal evolution of radio linear size could be interpreted in the light of observed D–z plane for radio source samples. Hence, the D–z correlation helps in constraining cosmological/temporal models. According to [50], the amount of linear size evolution required to interpret the observed θ–z data can be in the range of k = 2.0 and 1.5 to k = 1.2 and 0.75 for $\Omega = 1$ and 0 respectively. A clear investigation from the median values in **Figure 5** shows that the D–z curve of the observed data is in good agreement with that of the theoretical D–z plane of **Figure 3**. In other word, for $\Omega = 1$, there is an increase in the linear size of radio sources up to a certain point known as critical linear size, D_c which corresponds to critical luminosity, P_c at critical redshift, $z_c \sim 1$ and thereafter decreases. This steep change in P–z relation of extragalactic radio sources may occur around z = 0.3 [27, 33] or z = 1 (see also [46], Fig. 2) [51]. This is due to the luminosity selection effect occurred as a result of Malmquist bias in flux density limited samples [27]. In this effect, the present work adopted samples based on $z_c = 1$ as predicted earlier with the theory. This raise and fall of the radio sources in the D–z plane are consistent with the observational results showing that radio galaxies and quasars exhibit different D–P correlations.

In this scenario, the median values of D–P of the entire data superimposed in **Figure 6** clearly show that the observed D–P data of the sample are consistent with the theoretical curves (c.f. **Figure 3a**). Hence, this suggests that the P–D data of the present sample could be used to constrain the temporal evolution model.

A cursory look in **Table 1** shows that there is steep change in q as predicted above in the subsample around $z \geq 1$ with fairly trends with a significant correlation in the median values obtained from the eight appropriate luminosity bins of the entire sample. The results show that the apparent lack of D–P correlation in the samples below z = 1 was due to the luminosity selection effects in the observed data. In other words, in flux density limited samples, the low redshift ($z < 1$) sources exhibit more dependence on luminosity as a function of redshift than the high redshift ($z > 1$) source counterparts. Below z = 1, the P–z weak correlation is expected, while Beyond z = 1, the luminosities have a much milder dependence on z. The regression analyses yield q = +0.002 for all samples around $z \leq 1$ and q = -1.59 at z = ≥ 1. Hence, at $z < 1$, there is a positive temporal evolution model while at $z \geq 1$, negative temporal evolution is obtained for $z_c \sim 1$ and $\Omega_0 = 1$ as predicted. This shows that the temporal evolution model can be well understood using $\Omega_0 = 1$ than $\Omega_0 = 0$. In this scenario, the temporal evolution is positive when $z < 1$, $k > 0$ and $q > 0$, and otherwise for $z > 1$, $k < 0$ and $q < 0$. Thus, the linear size increases as a function of radio luminosity up to a maximum value called critical D-value, $Dc = 316kpc$ in consistence with the maximum theoretical linear size, $D_{max} = 0.14D_0$ at $z_c \sim 1$ and $P_c = 26.33WHz^{-1}$ and thereafter decreases with increasing luminosity, as predicted earlier in this work.

Several authors have argued that the unified scheme will vanish out if the derived positive D–P dependence for radio galaxies ($D \sim P^q$ for $q \sim 0.3$) contrasts with the negative D–P dependence in quasars [52]. Ubachukwu and Onuora [50] suggested an inverse correlation of D–P of the form $D \sim P^{-0.52}$. They found out that radio galaxies locate at low redshifts with $q \sim 0.3$ [31], while radio quasars located at high redshifts with $q \sim -0.5$ [48]. The D–P turnover in extragalactic radio sources occurs around Dc = 100kpc [29]. It was argued [24] that at $D \sim 1Mpc$ and critical luminosity, $P_c \sim 26WHz^{-1}$, extragalactic radio sources evolve and thereafter decrease.

In x–D regression analyses, there is an apparent lack of x–D anti-correlation in the entire samples and radio galaxies subsample. There was a fairly x–D anti-correlation in quasars subsample which was obvious in the upper envelope function with $r \sim -0.5$, 0.30 and $-$ 0.3 for the entire sample, radio galaxies and quasars respectively. Hence, the results suggest that orientation effects may be more necessary in explaining the properties of the radio quasars population than that of the radio galaxies populations in which the expected x–D anti-correlation was not observed and also implies that the two objects are the same, the difference is just the angle at which the observer took in viewing them.

3. Conclusions

The following major conclusions were drawn from the work. A satisfactory theoretical model was obtained to explain the observed D–P track, with sources at $z \geq 1$ showing stronger anti-correlation than those around $z < 1$. A problem was developed for these objects to match-up with D–z positive correlation as predicted in this paper at $z < 1$. Finally, the distributions of radio galaxies and quasars on the x–D plane are consistent with the beaming scenario, with quasars being more consistent with relativistic beaming and source orientation model than radio galaxies.

Author details

Costecia Ifeoma Onah[1*], Augustine A. Ubachukwu[2] and Finbarr C. Odo[2]

1 Department of Physics, Federal University of Technology, Owerri, Nigeria

2 Department of Physics and Astronomy, University of Nigeria, Nsukka, Nigeria

*Address all correspondence to: cossyonah@gmail.com

References

[1] Wang TL. Black hole mass and velocity dispersion of narrow line region in active galactic nuclei and narrow line Seyfert 1 galaxies. Astronomy & Astrophysics. 2001;**377**:52. DOI: 10.1051/0004-6361:20011071

[2] Urry CM, Padovani P. Unified schemes for radio-loud active galactic nuclei. Publications of the Astronomical Society of the Pacific. 1995;**107**:803

[3] Antonucci RRJ. Unified models for active galactic nuclei and quasars. Annual Review of Astronomy and Astrophysics. 1993;**31**:473. DOI: 10.1146/annurev.aa.31.090193.002353

[4] Kellermann KI, Sramek R, Schmidt m, Shaffer DB, and Green R. VLA observations of objects in the palomar bright quasar survey. Astronomy Journal. 1989;**98**:1195

[5] Kinwah W, Mark C, Gavin R, Kazuhiro S. An electrically powered binary star? Monthly Notices of the Royal Astronomical Society. 2002;**331**:221

[6] Garrington ST, Conway RG. The interpretation of asymmetric depolarization in extragalactic radio sources. Monthly Notices of the Royal Astronomical Society. 1991;**250**:198

[7] Zirbel E, Baum SA. On the FR I/FR II Dichotomy in powerful radio sources: Analysis of their emission-line and radio luminosities. Astrophysical Journal. 1995;**448**:521

[8] Ubachukwu AA, Chukwude AE. On the relativistic beaming and orientation effects in core-dominated quasars. Journal of Astrophysics & Astronomy. 2002;**23**:253. DOI: 10.1007/BF02702285

[9] Arshakian TG, Longair MS. An asymmetric relativistic model for classical double radio sources. Monthly Notices of the Royal Astronomical Society. 2000;**311**:846

[10] O'Dea CP. The compact steep-spectrum and gigahertz peaked-spectrum radio sources. Publications of the Astronomical Society of the Pacific. 1998;**110**:493

[11] Akujor CE, Spencer RE, Zhang FJ, Davis RJ, Browne WA, Fanti C. MERLIN observations of steep-spectrum radio sources at 6 cm. Monthly Notices of the Royal Astronomical Society. 1991;**250**:215

[12] Fanaroff BL, Riley JM. The morphology of extragalactic radio sources of high and low luminosity. Monthly Notices of the Royal Astronomical Society. 1974;**167**:31

[13] Owen FN, Laing RA. CCD surface photometry of radio galaxies -I. FR class I and II sources. Monthly Notices of the Royal Astronomical Society. 1989;**238**:357

[14] Prestage RM, Peacock A. The cluster environments of powerful radio galaxies. Monthly Notices of the Royal Astronomical Society. 1988;**230**:131

[15] Ferrari A. Modeling extragalactic jets. Annual Review of Astronomy and Astrophysics. 1998;**36**:539

[16] Miley GK. The structure of extended extragalactic radio sources. Annual Review of Astronomy and Astrophysics. 1980;**18**:165

[17] Perley RA, Bridle AH, Willis AG. High-resolution VLA observations of the

radio jet in NGC 6251. Astrophysical Journal (Supplement). 1984;**54**:291

[18] Kharb P, O'Dea CP, Baum SA, Daly RA, Mory MP, Donahue M, et al. A study of 13 powerful classical double radio galaxies. Astrophysical Journal Supplements. 2008;**174**:74

[19] Bridle AH, Hough DH, Lonsdale CJ, Burns JO, Laing RA. Deep VLA imaging of twelve extended 3CR Quasars. Astronomical Journal. 1994;**108**:776

[20] Longsdale CJ. In: Meisenheimer R, Roser HJ, editors. Hotspots in Radio Galaxies. Berlin: Springer; 1989. p. 27

[21] Burke BF, Graham-Smith F. An Introduction to Radio Astronomy. United Kingdom: Cambridge University Press; 1997

[22] Aird J, Wandra K, Laird ES, Georgakakis A, Ashby M. The evolution of the hard X-ray luminosity function of AGN. Monthly Notices of the Royal Astronomical Society. 2010;**401**:2531

[23] Miley GK. Variation of the angular sizes of quasars with red-shift. Nature. 1968;**218**:933

[24] Onah CI, Ubachukwu AA, Odo FC, Onuchukwu CC. Evolution of extragalactic radio sources and quasar/ galaxy unification. Revista Mexicana de Astronomia y Astrofisica. 2018;**54**:271

[25] Mattig W. Astronomische Nachrichten. 1959;**285**:1

[26] Onuora LI, Okoye SE. The variation of radio luminosity with epoch and its effect on the angular diameter-redshift relation. Astrophysical Journal. 1983;**270**: 360

[27] Ubachukwu AA, Ogwo JN. Luminosity selection effects and linear size evolution in the quasar/galaxy unification scheme. American Journal of Physics. 1998;**51**:143

[28] Gall C, Anderson AC, Hjorth J. Genesis and evolution of dust in galaxies in the early Universe I. Modelling dust evolution in starburst galaxies. Astronomy & Astrophysics. 2011;**528**: A13. DOI: 10.1051/0004-6361/ 201015286

[29] Okoye SE, Onuora LI. Angular diameter-redshift relations for extragalactic radio sources. Astrophysical Journal. 1982;**260**:37

[30] Ubachukwu AA. The implications of intrinsic luminosity evolution on the value of the density parameter and the evolution of radio sizes of radio galaxies and quasars. Astrophysics and Space Science. 1995;**228**:195

[31] Kapahi VK. Redshift and Luminosity dependence of the linear sizes of powerful radio galaxies. Astronomical Journal. 1989;**97**:1

[32] Peebles PJ. The theoretical aspects of the nebular redshift, 33 years later. Publications of the Astronomical Society of the Pacific. 1988;**100**:670

[33] Alhassan JA, Ubachukwu AA, Odo FC. On the absence of core luminosity-core-dominance parameter (P_C - R) correlation in radio galaxies and BL lacs. Journal of Astrophysics & Astronomy. 2013;**34**:61

[34] Fan JH and Zhang JS. JE The core dominance parameter of extragalactic radio sources. Astronomy & Astrophysics. 2003;**407**:899

[35] Bell MB and Comeau. Evidence that the bursting component of the X-ray radiation from 3C 111 originates in the

PC-scale jet. Astrophysics and Space Science. 2012;**339**:143

[36] Onuchukwu CC, Ubachukwu AA, Onah CI. On the expansion speed of highly asymmetric extragalactic radio sources. Nigerian Journal of Space Research. 2014;**12**:39

[37] Onah CI, Ubachukwu AA, Odo FC. Environmental effects on structural asymmetry in extragalactic radio sources and quasar/galaxy unification. European Physical Journal - Plus. 2021; **136**:576

[38] Ubachukwu AA. Relativistic beaming and orientation effects in lobe-dominated quasars. Astrophysics and Space Science. 1998;**257**:23

[39] Banhatti DG. Expansion speeds in extended extragalactic double radio sources from angular structure. Astronomy & Astrophysics. 1980;**84**:112

[40] Onah CI, Ubachukwu AA, Odo FC. Asymmetries in powerful radio sources and the quasar/galaxy unification. Journal of Astrophysics and Astronomy. 2014;**35**:619. DOI: 10.1007/s12036-014-9311-z

[41] Onuchukwu CC, Ubachukwu AA. Structural asymmetries, relativistic beaming and orientation effects in lobe-dominated quasars. Astrophysics and Space Science. 2013;**344**:211. DOI: 10.1007/s10509-012-1325-x

[42] Bai JM, Lee MG. New evidence for the unified scheme of bl lacertae objects and fr i radio galaxies. American Journal of Physics. 2001;**548**:244. DOI: 10.1086/318695

[43] Odo FC, Ubachukwu AA, Chukwude AE. On FRI-BL lac unification. African Skies/Cieux Africains. 2011;**15**:15-17

[44] Odo FC, Ubachukwu AA. Linear size-extended radio luminosity (D- P_E) correlation in BL Lacertae objects: evidence for large scale beaming? Astrophysics and Space Science. 2013; **347**:357. DOI: 10.1007/s10509-013-1528-9

[45] Vermeulem RC, Cohen MH. Superluminal motion statistics and cosmology, Astrophysical Journal. 1994; **279**:93

[46] Nilsson K. Kinematical models of double radio sources and the unified scheme II. The database. Astronomy & Astrophysics. 1998;**132**:31

[47] Landt H, Padovani P, Perlman ES. VLA observations of a new population of blazars. Astrophysical Journal. 2006;**637**: 183. DOI: 10.1086/498261

[48] Barthel PD, Miley GK. Evolution of radio structure in quasars: A new probe of protogalaxies? Nature. 1988;**333**:318. DOI: 10.1038/333319a0

[49] Gopal-Krishna and Wiita PJ. The expansion and cosmological evolution of powerful radio sources. Monthly Notices of the Royal Astronomical Society. 1987; **226**:531-542. DOI: 10.1093/mnras/226.3.531

[50] Ubachukwu AA, Onuora LI. Radio source orientation and the cosmological interpretation of the angular size–redshift relation. Astrophysics and Space Science. 1993;**209**:169-180

[51] Kollgard RI, Wardle JF, Roberts D, Gabuzda D. Astrophysical Journal. 1992; **104**:1687

[52] Singal AK. In: Ekers R et al., editors. Extragalactic Radio Sources, IAU Symp. 175. Kluwer: Dordrecht; 1996. p. 563

Chapter 3

The Unified Models for Black Hole Accretions

Xiaolong Yang

Abstract

Decades of observations and theoretical studies present intriguing results about black hole accretions: supermassive black holes (SMBHs), located in the centers of galaxies, are accreting similar to Galactic stellar-mass black hole systems (GBHs). This is the unified model of black hole accretion, which indicates active galactic nuclei (AGNs, the accreting SMBHs) are only the scaled-up version of galactic X-ray binaries (XRBs, the accreting GBHs). The analogy between AGNs and XRBs ensures us to determine AGN evolutions on cosmological timescales by simply studying the quick-playing Galactic systems, which is much easy in observation and modeling. X-ray emission is produced by the inner region of the accretion disk and corona, which is close to the black holes and provides the diagnostics of accretion strength. Meanwhile, radio emission is an indicator of the ejection process, which is another fundamental part of accreting black holes. Furthermore, accreting flows are also regulated by black hole masses and accretion rates/Eddington ratios. Therefore, the unified model of black hole accretion is the correlation between accretion and ejection process and black hole masses. In this chapter, we will review models concerning the unified model of black hole accretions and present recent updates in this area.

Keywords: accretion, stellar mass black holes, supermassive black holes, X-ray binaries, active galactic nuclei, jets, accretion disks

1. Introduction

Stellar-mass black holes, formed from the direct collapses of massive stars [1], are widely observed in the Universe. In contrast, supermassive black holes (SMBHs, $10^6 - 10^{10} M_\odot$) are also common in the centers of galaxies with bulges [2]. The accretion process is found in both stellar-mass black holes and supermassive black holes. Stellar-mass black holes accrete matters from a companion star and form X-ray binaries (XRBs). While, the accreting supermassive black holes at the centers of galaxies are observed as active galactic nuclei (AGNs), and they accrete matters from their host environment. Observations show that the structure of accretion flows around both XRBs and AGNs are similar and depend primarily on the accretion rates in terms of Eddington ratios. The accretion state transitions are associated with the evolution of Eddington ratios. With the evolution of Eddington ratios, the accretion flow or disk geometry will also change, meanwhile, resulting in multiband spectral features.

Galactic X-ray binaries (XRBs) can be well described with several distinct X-ray states, some of them being associated with jet launching [3]. A full evolution cycle of the state transition can be observed with convenient timescales (months to years), which was well explained as the evolution of accretion disk and jet-disk coupling [4]. It is now thought that the structure of accretion flows and jet production depends primarily on the Eddington ratio. As the Eddington ratio fluctuates, the accretion flow transitions dramatically into different states, each with distinct geometries and multiwavelength spectral characteristics [5]. The current observational picture of state/disk-jet correlation is: (a) in the "hard" state, which exists typically below a few percent of the Eddington luminosity, there is a compact and steady jet; (b) subsequently, the transition from "hard" to "soft" state always associated with a transient/episodic jet, which corresponds to a "very high state" with near/super-Eddington rates; (b) in steady "soft" states with Eddington ratio lower than the very high state, the jet production is strongly suppressed. It was noted that with the accretion rate increasing to near and moderately super-Eddington ratios, the standard disk cannot maintain its geometry and will inevitably evolve into a "slim disk" [6], the corresponding state in observation was named as "ultraluminous state" [7]. The study of jet-disk coupling in "ultraluminous state" is limited to a few XRBs that can temporarily transit to super-Eddington accretion and the long-lived super-Eddington source SS 433.

The theoretical understanding of the state transition is explained as the evolution of the accretion disk. **Figure 1** shows the geometry of the disk in different accretion states. The quiescent state XRBs host low accretion flow with Eddington ratio $\lambda_{Edd} < 0.01$ [9]. The accretion flow in a quiescent state can be described as advection-dominated accretion flows (ADAFs, Narayan and Yi [10]). The advection-dominated accretion flows are radiatively inefficient (Shapiro et al. 1976). With the increase in accretion rates, the accretion flow of X-ray binaries transit to a geometrically thin accretion disk, i.e., the so-called standard accretion disk or Shakura-Sunyaev disk [11]. During this state, the X-ray spectrum becomes dominated by comptonized hard

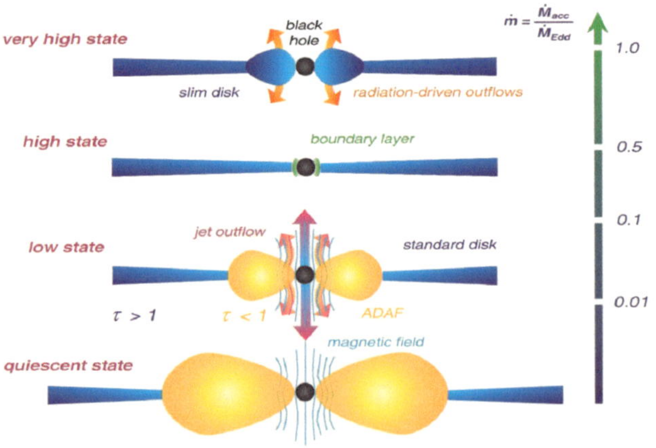

Figure 1.
Illustration of the spectral states of black hole accretion disks from [8]. The accretion rate is given in terms of the Eddington ratios.

X-rays. With further increase of Eddington ratio, the accretion flow becomes hot and luminous, it emits soft-X-ray emission with a thermal spectrum, the spectra are characteristics as soft state. Between the low/hard state and high/soft state, there is sometimes an intermediate state, corresponding to an unstable accretion flow. In the intermediate state, the accretion will have an extremely high or super-Eddington accretion rate, its accretion disk is described as a slim disk. For this reason, the intermediate state is also called the very high state. Despite the success of this general picture for accretion state transitions in stellar-mass black holes, it remains unclear if supermassive black hole accretion flows undergo similar processes.

Several schemes are successful in unifying black hole accretion flows in active galactic nuclei (AGNs) and Galactic X-ray binaries (XRBs) [12–14], it is now widely accepted that supermassive and stellar-mass black holes have similar physics in accretion, i.e., AGNs and XRBs have similar accretion states and associated ejection (especially in low/hard state). Over several years, observations have built kinds of universal correlation between XRBs and AGNs: (1) the fundamental plane of black hole activity reveals a correlation among radio luminosities, X-ray luminosities, and black hole masses [12, 15]. The correlation can be well applied to both low and moderate accretion rates (in Eddington units) XRBs and AGNs. The fundamental plane correlation of black hole activity suggests that both the accretion and ejection process are regulated by black hole masses; (2) similarly, a more universal correlation is found between radio loudness and the Eddington ratio, which hints at the suppression of the ejection process with the increase of accretion rates in units of black hole masses [13, 16–18]. The correlation has a broader application as it covers from low to super-Eddington rates; (3) another fundamental correlation of black hole accretion is among the characteristic timescales of X-ray variability, bolometric luminosities, and black hole masses [19]. The correlation links the accretion process and black hole mass in both XRBs and AGNs, which indicates accreting black holes have mass regulated disk geometry; (4) The most fruitful result in studying accretion states and transitions in XRBs is the hardness intensity diagram, while in applying the scheme to AGNs, it has big problem primarily due to the extremely long timescale in evolution cycle of AGNs. Therefore, the disk-fraction luminosity diagrams [14, 20] are taken as an alternative scheme in AGNs.

However, none of the above correlations are applicable to all accretion states or Eddington ratios. Furthermore, some extreme accretion states, for example, the extremely low accretion flow, the very high/intermediate state, and the super-Eddington state, are not fully understood in studying XRBs. Especially the models for the ultraluminous/super-Eddington state are not established yet due to the short timescales in XRBs. Furthermore, for example, it's not clear whether the accretion of intermediate-mass black holes can follow the fundamental plane of black hole activity. It is thus questionable when applying the fundamental plane of black hole activity to constrain the black hole mass of AGNs in dwarf galaxies.

2. The universal correlations among accretion systems

2.1 X-ray variability

X-ray emission, produced from the inner region of the accretion disk and corona, served as a proxy of accretion properties. X-ray emissions from accreting black holes have strong variability, the timing properties of X-ray emission can be explored with the power spectral densities (PSDs), $P(\nu)\propto\nu^{-\alpha}$, which is a function of timescales $1/\nu$ of the variability. In both XRBs and AGNs, the PSD has a power-law spectrum. The spectral index $\alpha\approx1$ on long timescales, while it is transient to a steep spectral index $\alpha > 2$ on short timescales. The characteristic timescale, $T_{\rm B}$, on where the PSD break or transitions, is a common feature in both XRBs and AGNs. Using the timescale, $T_{\rm B}$, as a representation of black hole accretion was established with the finding of a correlation between $T_{\rm B}$ and black hole mass M_{BH} was established by [19].

Again, it was strengthened as the break timescale is also correlated with spectral states or luminosities of both XRBs and AGNs, i.e., the low and high accretion states have different PSD profiles. Therefore, $T_{\rm B}$ is a good reflector of black hole masses and accretion states (in terms of bolometric luminosity T_{bol}). The correlation among the critical timescale $T_{\rm B}$, black hole mass M_{BH}, and bolometric luminosity L_{bol} is fitted by [19] (see **Figure 2**) as Eq. (1)

$$\log T_B = (2.10 \pm 0.15)\log M_{\rm BH} - (0.98 \pm 0.15)\log L_{bol} - (2.32 \pm 0.2) \qquad (1)$$

They have included 10 AGNs and 2 XRBs and with a wide range of accretion rates.

Assuming $\dot{m}_{\rm E}\approx\lambda_{\rm Edd}$, then $T_{\rm B}\approx M_{\rm BH}^{1.12}/\dot{m}_{\rm E}^{0.98}$, where $\lambda_{\rm Edd} = L_{\rm bol}/L_{\rm E}$. If the break timescale is proportional to a thermal or viscous timescale associated with the inner radius of the accretion disk, $R_{\rm disk}$, then from the above empirical correlation, it can be deduced $R_{\rm disk}\propto\dot{m}_{\rm E}^{-2/3}$. Models based on evaporation of the inner accretion disk predict $R_{\rm disk}\propto\dot{m}_{\rm E}^{-0.85}$ and $T_{\rm B}\propto M_{\rm BH}^{1.2}$, which are roughly consistent with the empirical correlation.

Strong support or enhancement for this linkage, among characteristic timescale $T_{\rm B}$, black hole masses, and luminosities, comes from the correlation between $T_{\rm B}$ and emission line region in the circumnuclear region of black holes (see **Figure 3**). As the emission line width, the full width at the half maximum *FWHM* is regulated by black hole masses and accretion rates. Therefore, the correlation between *FWHM*

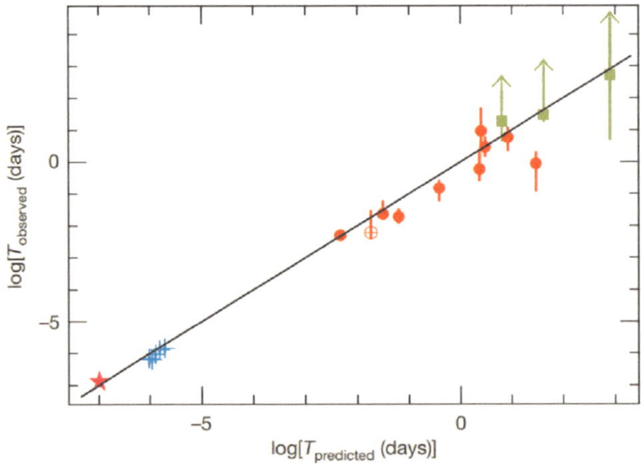

Figure 2.
The predicted break timescales, $T_{predicted}$, derived by inserting the observed bolometric luminosities and masses into the best fit relationship (Eq. 1) to the combined sample of AGNs and GBHs [19]. Where GRS 1915+1105 is presented as a filled maroon star, Cyg X-1 as blue crosses, and the 10 AGN as red circles. The low-luminosity AGN (LLAGN), NGC 4395, is shown as an open crossed red circle; the other nine AGN are filled red circles. The filled green squares are NGC 5548, and Fairall 9 and the LLAGN NGC 4258.

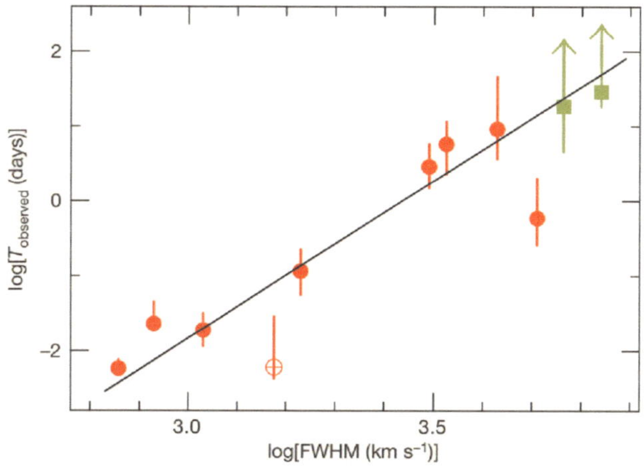

Figure 3.
Correlation of optical emission linewidth FWHM with PSD break timescale T_B from [19]. The LLAGN NGC 4395 is shown as an open crossed red circle and the other eight AGNs are filled red circles. Filled green squares are Fairall 9 and NGC 5548, their linewidths are available but whose upper break timescales are only lower limits.

and T_B is a straightforward product. The correlation is explored as the permitted optical emission lines in AGN whose widths (in both broad-line AGN and narrow-emission-line Seyfert 1 galaxies) correlate strongly with the characteristic X-ray timescale (see **Figure 3**),

$$\log T_B = \left(4.20^{+0.71}_{-0.56}\right) \log FWHM - 14.3 \tag{2}$$

2.2 The fundamental plane of black hole activity

The fundamental plane relation among nuclear radio luminosity, nuclear X-ray luminosity, and black hole mass unified the accretion and ejection process in the compact system. The existence of such a relationship is based on the radio emission produced in a jet/outflow and the X-ray emission produced in a disk-corona system. Both radio and X-ray power are related to black hole mass and accretion rate. Therefore, the fundamental plane relation is thought to work in any accretion system, which is in quiescent and low/hard accretion state (associated with a steady ejection, see [9, 12]).

The fundamental plane of black hole activity explored by [12] is

$$\log L_R = \left(0.60^{+0.11}_{-0.11}\right) \log L_X + \left(0.78^{+0.11}_{-0.09}\right) \log M_{\rm BH} + 7.33^{+4.05}_{-4.07}, \tag{3}$$

While the Merloni's fundamental plane has a large dispersion when uses it to estimate black hole masses from radio and X-ray luminosities. The recent updates include correlation from [21]

$$\log L_R = \left(0.60^{+0.11}_{-0.11}\right) \log L_X + \left(0.78^{+0.11}_{-0.09}\right) \log M_{\rm BH} + 7.33^{+4.05}_{-4.07}, \tag{4}$$

and the most recent version [22]

$$\log\left(\frac{M_{\rm BH}}{10^8 M_\odot}\right) = (1.09\pm0.10)\log\left(\frac{L_R}{10^{38}\,{\rm erg\,s^{-1}}}\right) + (-0.59^{+0.16}_{-0.15})\log\left(\frac{L_X}{10^{40}\,{\rm erg\,s^{-1}}}\right) + (0.55\pm0.22) \quad (5)$$

Additionally, the very high/intermediate state may also produce radio ejection that can follow the same trend [see 4, they also include transient sources]. However, including sources with a very high/intermediate state induces a dispersion in the fundamental plane relation. This is primarily due to the evolution of individual radio blobs, as the radio ejecting process is episodic in this state. Compact symmetric objects (CSOs) are thought the episodic ejection produced by AGNs in a very high/transient state. There are two types of known contamination in the fundamental plane of black hole activity: (1) radio emissions from lobes will be enhanced when they propagate through a dense medium [23]; (2) X-ray emission contains a contribution from the jet, e.g., through synchrotron or inverse Compton mechanisms [24]. Furthermore, taking the radio emissions from lobes of CSOs is unmatched by the X-ray observations, because the radio emissions from lobes are substantially produced in different epochs from the core X-ray emission.

There are several works exploring the fundamental plane relation on CSOs [25–27]. Most of the results suggest CSOs do deviate from the classical trends. To be specific, an exploration of fundamental plane relation on a sample of CSOs (with radio flux density from lobes) indicates that they can follow the trend, while their radio luminosity is $\sim$1 dex higher than the original fitting of the fundamental plane relation [see 26], which is consistent with the transient state in XRBs. Another point is the vacuum region between stellar-mass black holes and supermassive black holes in the fundamental plane of black hole activity. This region can be filled by either intermediate-mass black holes or extremely low luminosity AGNs. We will discuss it in the following sections.

2.3 The inverse correlation between radio loudness and Eddington ratio

The persistent jets are ubiquitous at low accretion rates (the low/hard state) in XRBs but intermittent or entirely absent at high accretion rates (the high/soft state and the intermittent/very high state; e.g., [3, 28, 29]). Resembling the inverse correlation between the radio luminosity of jets and X-ray luminosity in XRBs, Ho [16] found a similar inverse correlation between radio-loudness ($\mathscr{R} = L_{\nu 5}/L_{\nu B}$) and the Eddington ratio ($\lambda_{Edd} = L_{bol}/L_{Edd}$) in AGNs. In this scheme, radio-loud AGNs with powerful relativistic jets often have low Eddington ratios and vice versa. In contrast to the fundamental plane relation only being valid for certain conditions, the inverse correlation between $\mathscr{R}$ and λ_{Edd} is ubiquitous in both AGNs and XRBs [18] although with a large scatter. A global analogy between stellar-mass black holes and SMBHs has been established in the $\mathscr{R}$ and λ_{Edd} correlation: low-luminosity AGNs are similar to XRBs in the low/hard state, and with the high- or super-Eddington accreting AGNs (e.g., NLS1s) being an analogy of XRBs in the high/soft and the very high state. It should be noted here that only a few XRBs experience transitions from classical spectral states to the super-Eddington regime [30]. The super-Eddington accretion state is poorly understood primarily because of the extremely short timescales of the

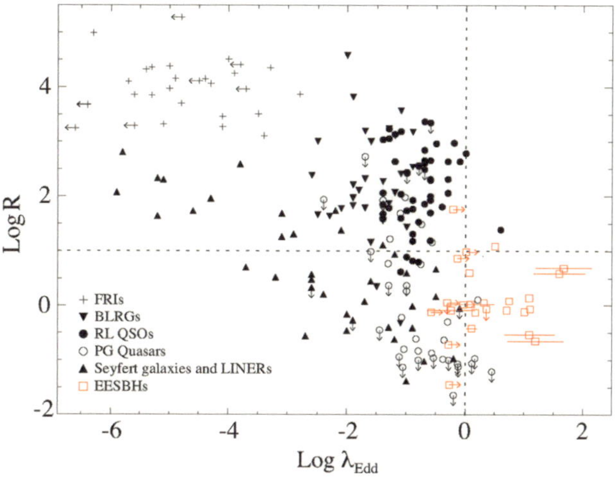

Figure 4.
Radio loudness $\mathscr{R}$ vs Eddington ratio λ_{Edd}. The markers are designated in the left-bottom corner and the error bars in some extremely high Eddington ratio accreting SMBHs come from the uncertainty of the BH spin. The vertical dotted line is $\lambda_{Edd} = 1$ and the horizontal dotted line at $\mathscr{R} = 10$ represents the division between radio-loud (above) and radio-quiet (below) sources.

spectral state transition in XRBs. Therefore, the study of the production and quenching of AGN jets in super-Eddington accretion systems will shed light on the physical properties of AGNs during this short-lived spectral state (**Figure 4**).

In the work by Yang *et al.* [13], the inverse correlation between radio loudness and the Eddington ratio has been extended to the super-Eddington regime. It was shown that the correlation is even more fundamental than between radio and X-ray luminosity and black hole mass, as the correlation works among extremely low-power accreting black holes. However, the radio loudness-Eddington ratio correlation has a large dispersion, and it suggests there is an additional parameter at work.

2.4 X-ray loudness versus Eddington ratio

It is impossible to observe a whole state transition in AGNs due to their extremely long evolutionary timescales. While, fortunately, an unbiased sample of AGNs will naturally have a mixture of AGNs in various accretion states, e.g., an AGN sample includes low-luminosity AGNs (LLAGNs), low-excitation emission line regions (LINERs), and narrow line Seyfert I galaxies (NLS1s). The properties/structures of the accretion disk and corona are represented by the X-ray loudness or UV to X-ray spectral index α_{ox}, which is defined, for example [31], as

$$\alpha_{\mathrm{ox}} = -\frac{\log(\nu L_\nu)_{\mathrm{o}} - \log(\nu L_\nu)_{\mathrm{x}}}{\log\nu_{\mathrm{o}} - \log\nu_{\mathrm{x}}} + 1, \tag{6}$$

Where $(\nu L_\nu)_{\mathrm{o}}$ and $(\nu L_\nu)_{\mathrm{x}}$ are monochromatic immensities at the rest-frame optical/UV and X-ray energies, $\lambda_{\mathrm{o}} = c/\nu_{\mathrm{o}} = 2500\text{Å}$ and $E_{\mathrm{x}} = h\nu_{\mathrm{x}} = 2\mathrm{keV}$, respectively; correlates primarily with $L_{\nu,x}$; and that there is a strong correlation between $L_{\nu,\mathrm{o}}$ and $L_{\nu,\mathrm{x}}$ (**Figure 5**).

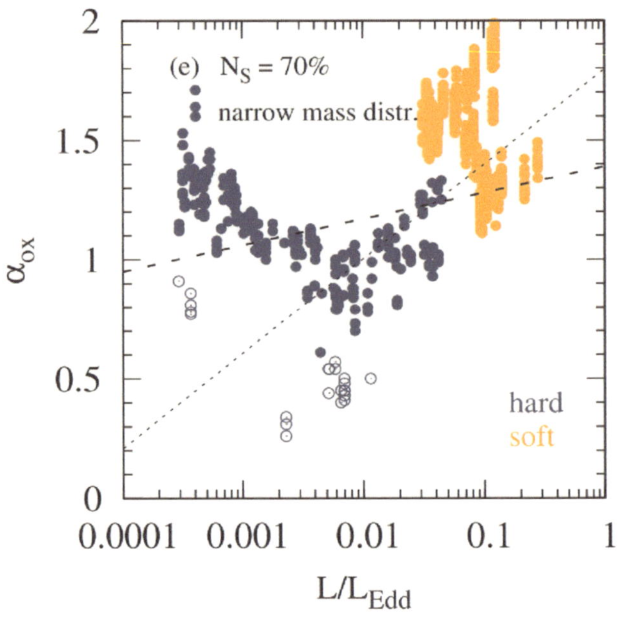

Figure 5.
The transition behavior along X-ray loudness α_{ox} versus Eddington ratio L/L_{Edd} [31]. A change of the sign of the α_{ox} can be observed for $L/L_{Edd} = 0.01$. The transition behavior is free of black hole masses. The dotted and dashed lines correspond to the correlations found by Lusso et al. [32] and Grupe et al. [33], respectively.

A characteristic spectral behavior was found by taking the typical galactic X-ray binary GRO J1655−40 as a template, which drives a complete state transition within 1 year. The X-ray loudness versus Eddington ratio distribution has a "V"-shaped morphology, i.e., in low Eddington ratios of $L_{bol}/L_{Edd} < 0.01$, the X-ray loudness has an inverse correlation with Eddington ratio, while in higher Eddington ratios of $L_{bol}/L_{Edd} > 0.01$, XRBs show a positive correlation between X-ray loudness and Eddington ratio.

The correlation between X-ray loudness and Eddington ratio can be explained as the evolution of accretion flow along with accretion state transition. In the quiescent state, $L_{bol}/L_{Edd} < 0.01$, the accretion disk is described as the radiatively inefficient advection-dominated accretion flow (ADAF) and drives a compact jet, where ADAF dominates the UV emission and corona dominate X-ray emission. Furthermore, the jet may also contribute to UV and X-ray emissions in this state. With the increase of Eddington ratios in this state, (1) the inner ADAF will progressively fall to touch with the outer thin disk (SSD), leading to a decrease in UV emission; (2) the corona will extend upward from the accretion disk and enhance the X-ray emission. Eventually, the increase of the Eddington ratio in this state leads to the hardening of the UV to X-ray spectrum. With the further increase of Eddington ratios from $L_{bol}/L_{Edd} = 0.01$ to 1, the hot accretion flow of the thin disk produces strong thermal UV emission and the corona will be suppressed into small scale. This leads to a signature that with the increase of Eddington ratios the UV to X-ray spectrum will gradually soften and result in the positive correlation between α_{ox} and L_{bol}/L_{Edd}.

It should be expected that XRBs and AGNs have a similar accretion flow and evolution scheme, which corresponds to the straightforward XRB/AGN analogy. The correlation between α_{ox} and L_{bol}/L_{Edd} was found in a large unbiased sample of AGNs [34–40], indicating that the different types of AGNs have an evolutionary connection with each other and the AGN accretion flow may evolve similarly with XRBs. Recently, a fascinating finding is that the changing-look AGNs evolve along the α_{ox} vs. L_{bol}/L_{Edd} trend, which hints at the state transition in individual AGN.

3. In extreme cases

3.1 Extremely high and super-Eddington accretion

Accretion of black holes at near-Eddington or super-Eddington rates is the most powerful episode in nursing black hole growth [41], and it may work in several types of objects [13, 42–51]. It is still unclear whether the AGN/XRB analogy holds in the "ultraluminous state," and whether the geometry of the disk-corona system and jet-disk coupling are similar. Since it is impossible to observe the whole burst cycle of an individual AGN as the timescale is proportional to the black hole mass [13, 14], the previous studies rely on a large enough unbiased sample of AGNs, which naturally contains a mixture of objects in different spectral states. While near/super-Eddington AGNs provide an opportunity that it has a longer timescale than the short-lived "ultraluminous state" in XRBs and potentially connect with long-lived super-Eddington sources (SS 433 and ultraluminous X-ray sources). On the other hand, the jet is a long-lived emitter that saved long timescale information of an accretion state. Observationally confirming jet properties of the less explored "ultraluminous state" in AGNs would enhance the AGN/XRB analogy, as it enables us to eventually apply our understanding of X-ray binaries to explain AGN phenomenology (and vice versa). On the other hand, the study of near/super Eddington AGNs will shed light on our understanding of the physics to sustain a near/super-Eddington accretion and how the episodic jet works in this state.

The physics of accretion and jet-disk coupling in such a state remains unclear [52], mainly because the associated jets are not easily detectable due to the extremely weak or episodic nature of the jets. Although a few near/super Eddington objects have demonstrated jet activity [42–44, 46–49, 51, 53, 54], in most of these systems, such as super-Eddington active galactic nuclei (AGNs) [13] and ultraluminous X-ray sources [45], it remains doubtful whether there is jet emission. Recent observations [13, 55–58] suggest that the radio emission in near/super-Eddington AGNs comes from the nuclear region, with possible contributions from hot corona, accretion disk winds, fossil radio jet, or a combination of all above [13, 55, 58–62]. In particular, unambiguous detection of both powerful radio jets and radio-emitting winds was reported only in the Galactic microquasar SS 433 [63–65], while accretion disk wind is believed to be ubiquitous in the near/super-Eddington accretion mode.

It is now thought that the structure of the accretion flows and jet production depends primarily on the Eddington ratio. As the Eddington ratio fluctuates, the accretion flow transitions dramatically into different states, each with distinct geometries and multiwavelength spectral characteristics [5]. As the accretion rate increases to near or super-Eddington ratios, the standard disk geometry cannot be maintained and the accretion flow will inevitably evolve into a "slim disk" [6]. The corresponding state is sometimes called the "ultraluminous state" [7]. Studies of jet-disk coupling in "ultraluminous state" have been limited to a few XRBs that can temporarily transit to super-Eddington accretion and to the long-lived super-Eddington source SS 433. It is also widely accepted that supermassive and stellar-mass black holes have similarities in accretion physics, i.e., AGNs and XRBs have similar accretion state transitions and associated jet ejection. However, it is still unclear whether the AGN/XRB analogy holds in the "ultraluminous state" and whether the geometry of the disk-corona system and jet-disk coupling are similar. Here, our interest is the connection between the short-lived canonical "very high state" in XRBs with the long-standing super-Eddington accretion in the microquasar SS 433 and ULXs, to determine which parameters are driving the long-lived super-Eddington accretion. As the time scale of state transition is proportional to the black hole mass [13, 14], a "very high state" in SMBHs (e.g., $M_{BH} = 10^7 M_\odot$) would last 10^6 times longer than in $10 M_\odot$ stellar-mass black holes found in XRBs. Therefore, the study of near/super-Eddington AGNs provides an opportunity to understand the ejection process in a quasi-steady "very high state" and may shed light on the physics to sustain a near/super-Eddington accretion.

3.2 Intermediate mass black holes

Directly connecting stellar-mass and supermassive black holes requires intermediate-mass black holes [66]. In the unified model of black hole accretion, filling intermediate-mass black holes will build a continuous distribution of accretion parameters. It is now widely accepted that the existence (or not) of intermediate-mass black holes (IMBHs, $M_{BH} = 10^2 - 10^6 M_\odot$) is an even more fundamental question, and it has an essential impact on our theoretical deduction of black hole formation and evolution [see 18, and references therein]. It is believed that stellar-mass black holes are formed from the direct collapse of massive stars [1]. Such black holes are known to be abundant in our Galaxy. On the other hand, supermassive black holes (SMBHs, $M_{BH} = 10^6 - 10^{10} M_\odot$) are universally found in the centers of massive galaxies with bulges [2]. Mergers and accretion are known as the primary and effective ways to drive black hole growth. Observations indicate SMBHs with masses up to $10^{10} M_\odot$ [67, 68] have already existed when the Universe was only 5% of its current age. However, to assemble SMBHs through accretion would require dramatic feeding, which poses a severe challenge to the formation of SMBHs [see 18]. Seed black holes with intermediate-mass [IMBHs, $M_{BH} = 10^2 - 10^6 M_\odot$ are needed in the very early Universe when the first-generation SMBHs have not formed.

Astrophysical black holes (BHs), inferred through their observational signatures (electromagnetic, gravitational waves), are currently understood to fall into two categories based on their mass. Stellar-mass BHs ($3 - 100\ M_\odot$) originate from the end stages of the evolution of massive stars, as has been inferred from studies of X-ray binaries (BH actively accreting from a companion star) in our Galaxy. Supermassive BHs (SMBHs; $\geq 10^6\ M_\odot$) on the other hand are resident at the centers of most massive galaxies. These have been mainly inferred through their role in the evolution of the host galaxy (through the correlations of the SMBH mass with the galactic bulge properties, including the dispersion velocity, luminosity, and mass). As there have been deductions of SMBH hosts even in the early Universe (less than a Gyr) through their observational signatures (accretion power and nuclear activity), modes of growth to such large masses ($10^6 - 10^{10}\ M_\odot$) remain debatable. Possibilities include mergers and accretion activity. These scenarios require a rapid progression involving lower mass seed BHs, which are plausible. The presence of intermediate-mass BHs (IMBHs; $10^2 - 10^6\ M_\odot$) can help realize these scenarios more efficiently than lower mass seed BHs.

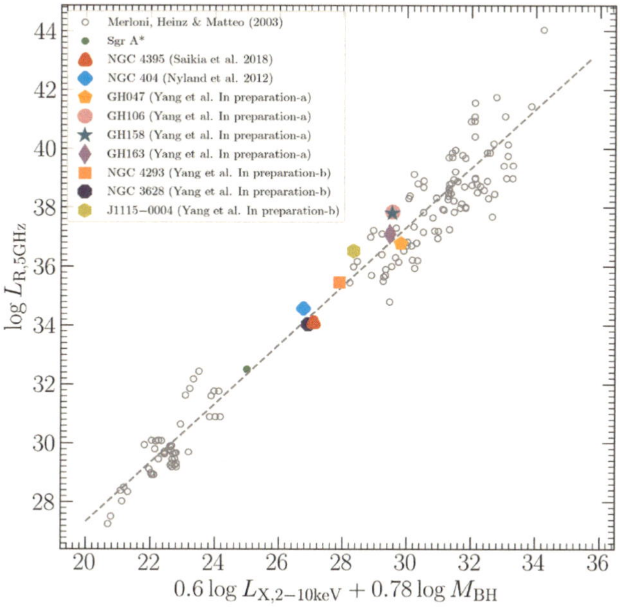

Figure 6.
The fundamental plane relation of black hole activity based on [12]. The references in the legend show where the radio luminosity was taken. The black open squares and data for Sgr A are from Merloni, Heinz, and di Matteo [12]. Note that the radio luminosities for NGC 3628, NGC 4293, and J1115−0004 are only upper limits.*

IMBHs can follow the fundamental plane of black hole activity. This indicates that an outflow-disk-corona system still exists and is tightly related even in these systems. In **Figure 6**, we also include two well-studied IMBHs NGC 4395 and NGC 404. Especially, we take recently measured black hole mass of NGC 4395 [69] and NGC 404 [70]. Again, we take radio luminosity of NGC 404 from VLA A-array 5 GHz observation [71], which captured radio emission from the nuclear 7 parsec scale region; we take 5 GHz radio luminosity of NGC 4395 transferred from VLA A-array 15 GHz observation [21], which captured radio emission from the nuclear 4 parsec scale region. Because of low redshift, the VLA observation of NGC 4395 and NGC 404 obtained a resolution of parsec scale, which is comparable with VLBI observations on slightly high red-shift AGNs. We take the X-ray luminosity of NGC 404 from *Chandra* observation [72].

3.3 Low-luminosity AGNs

In addition to the requirement of IMBHs, low-luminosity AGNs will touch the most luminous stellar-mass black holes in the fundamental plane of black hole activity. In quiescent state XRBs, there exist compact radio emissions and are thought to be short and steady jets, while the nature is unclear. However, in low-luminosity AGNs, radio emissions from the core region are consistent with wind-like outflows or low-power jets. Synchrotron emission as the result of propagating shocks produced and sustained by the injection of new material at the base of the outflow accelerated electrons downstream to relativistic energies. Low luminosity sub-Eddington emitting sources could host advection dominated accretion flows [ADAFs, 73] that are radiatively inefficient in the inner region [e.g., 16]. This can include nearby dwarf galaxies (low mass low luminosity systems) hosting an inner truncated region, with the outer thin disk accretion [optically thick, geometrically thin, e.g., 31] transitioning into an ADAF [74]. Radio emission in these systems can be contributed to by the ADAF but is likely to be dominated by the jet or outflow [75–77], with observable signatures including shock ionization of the gas in the nuclear region [e.g., 78, 79].

It was shown that such low accretion flow deviates from the plane [80]. While in exploring the universal correlation between XRBs and AGNs, one should obtain radio emissions from the same radius, i.e., with regard to the Schwarzschild radius. Therefore, a moderate resolution is enough in a few nearby AGNs. M32 is one of the prominent low-luminosity AGNs with an Eddington ratio of only $\sim 10^{-8.5}$ [81]. The X-ray emission of M32's AGN is detected by *Chandra* [81]. The radio luminosity was obtained from VLA B-array 6.6 GHz [82] and VLA A-array 6 GHz [83] observations. The two VLA observations of M32 obtained a resolution of $\sim$ 4 and $\sim$ 1.5 parsec, respectively. Again, the IMBHs NGC 404 and NGC 4395 are both low-luminosity AGNs, they have Eddington ratio $\lambda_{\mathrm{Edd}} = 1.5 \times 10^{-6}$ [72] and 1.2×10^{-3} [84], respectively. We note that these low-luminosity and low (or intermediate)-mass AGNs tend to have steep radio spectra and diffuse radio emissions. The radio emissions fall below the detection threshold with the resolution higher than $\sim$ 1 parsec scale (Yang et al. In preparation) [72, 83], which results in underestimation of radio luminosity in the fundamental plane. Therefore, the radio emission can be explained as wind-like outflows driven by weakly accreting AGNs [85]. As the fundamental plane of black hole activity looks reliable for most low-mass AGNs, which suggests that a moderate resolution, as well as high sensitivity, should be taken to fully collect wind-like radio emission produced by the central engine but avoid contamination from hosts, i.e., between $\sim$1 and $\sim$10 parsec scale region. In exploring the fundamental plane relation of black hole activity for both XRBs and AGNs, it is reasonable to constrain radio emission from a similar region with regard to Schwarzschild radius. Meanwhile, it's still possible that low-luminosity AGNs deviate from the fundamental plane relation [e.g., 80].

3.4 Capturing the state transition in AGNs

Changing-look AGNs (CLAGNs) are a subclass of AGNs, they change the spectral type from type 1 to type 2 (disappearance of the broad emission line) or vice versa (emergence of the broad emission line) on timescales shorter than a few years [86]. The spectral-type changes in CLAGNs are commonly associated with multiband continuum behaviors [86]. The changing look of AGNs challenges the unified model of AGN [87, 88]; however, it provides a chance to explore the dramatic state transition in AGNs.

Directly capturing the changing-look events when it is in the act is essential to explore the accretion state transition in AGNs. The chance comes from 2018, a rapid spectral-type change was observed in the Seyfert 2 AGN 1ES 1927 + 654 ($z = 0.017$), which was followed up with multiband observations, including in the X-ray, optical, and radio wavelengths. The All-Sky Automated Survey for SuperNovae (ASAS-SN) first reported an optical flare from the nuclear region of 1ES 1927 + 654 on 2018-2103-03 [ATel #11391, 89]; this was accompanied by the emergence of broad Balmer lines in the optical spectrum [90] with the consequent classification as a changing-look AGN (from Type 2 to Type 1). The Neutron star Interior Composition Explorer (NICER) observations of 1ES 1927 + 654 (on 2018-2105-22) found an extremely soft

X-ray spectrum and a continued decrease in the X-ray luminosity [ATel #12169, 91] compared with archival data. This is followed by the NICER detection of an increase in the X-ray luminosity beyond 1st July [ATel #12169, 91], 4 months after the optical outburst. The dense optical/UV and X-ray monitoring observations [90, 92] confirm the changing-look nature of 1ES 1927 + 654.

1ES 1927 + 654 has been reported to show unusual timing and spectroscopic properties. The nuclear region is relatively unobscured based on a low neutral gas column density from X-ray observations (lack of sufficient absorbing gas along the line of sight); this and timing properties are reminiscent of a Seyfert type 1 [93]. However, optical spectroscopic observations reveal a Seyfert type 2 nuclear region [93, 94]. These pose challenges for the line-of-sight-based AGN unification model [e.g., 63]. A previous lack of broad optical emission lines typical of Seyfert type 2 galaxies with their prominent appearance post the changing-look event [90], accompanied by a relatively unobscured X-ray emission [95], suggests an origin (of the emission lines and the changing-look event) associated with physical processes in the accretion flow. The studies of [92, 95] find an X-ray spectrum dominated by the soft (black-body, disk) continuum with the disappearance of the hard power-law component following the optical/UV outburst. The disappearance and subsequent reappearance of the power-law component (with an accompanying increase in luminosity) are interpreted as the destruction and recreation of the accretion disk. One of the promising models for the changing look in 1ES 1927 + 654 is the consequent evolution of the jet/outflow and radiative properties [96].

The radio emission can originate from an outflow (collimated/relativistic or wide-angled/nonrelativistic). Propagating shocks either internal to the outflow [injection events from accretion – outflow activity, e.g., 97] or as a consequence of its interaction with the surrounding medium [e.g., 98] can accelerate electrons downstream with the consequent emission of synchrotron radiation. 1ES 1927 + 654 has been studied in the radio bands, with successful VLBI observations conducted in epochs prior to, covering, and post the changing-look event. Very long baseline interferometric observations of 1ES 1927 + 654 revealed exciting results, which provide further constraint on the quick accretion state changing in this source [99]: (1) The European VLBI Network (EVN) observation during 2013–2014 yields a radio to X-ray luminosity ratio $\sim 10^{-5}$ and a steep radio spectrum, suggesting that the radio emission likely originates from an outflow; (2) a long-time decline in radio flux density is similar to that in the optical and X-rays, which confirms a multiband decay over past $\sim$ 30 years; (3) recently, we have successfully detected an increase of radio flux density, which is $\sim$ 700 and $\sim$ 450 days delayed since the optical and X-ray flare (Yang et al. ATel), respectively; (4) from the VLBA X-band observation in 2020, we have resolved for the first time the innermost structure of this source. A continued monitoring observation of radio emission is still ongoing, we are expecting to see further intriguing evidence to constrain properties of the outflow (proper motion, radiative evolution, and association with the accretion, total energy, and magnetic field strength) and surrounding environment (number density, density contrast).

4. Conclusion and future directions

Accretion is an essential process to drive black hole growth, and it is thought to work in different types of accreting black hole systems from stellar-mass galactic black holes to supermassive black holes located in the centers of galaxies. Now, we are near to reaching a consensus that the physics in controlling the accretion and the associated ejecting process is exactly same in various kinds of accreting systems. In this chapter, we explored the correlations concerning the universal evolution among various accreting systems: (1) the correlation among X-ray variability, the black hole mass, and bolometric luminosity; (2) the fundamental plane of black hole activity, i.e., the correlation among core X-ray and radio luminosity and black hole mass; (3) the inverse correlation between radio loudness and Eddington ratio; (4) the correlation between X-ray loudness and Eddington ratio. These evidences ensure us to apply the theory to study the accretion process in, e.g., high red-shift quasars, the evolutionary connection between FRI and FRII radio galaxies, and the accretion signatures of low luminosity AGNs, and so on. However, the unified models for black hole accretions still face challenges in a few types of cases in practical applications: (1) the extremely high and super-Eddington accreting systems are poorly understood in both XRBs and AGNs; (2) the intermediate region between stellar-mass black holes and supermassive black holes in the fundamental plane of black hole activity is still unfilled; (3) the intermediate-mass black holes are absent; (4) the lack of evidence of state transition in individual AGNs still throws doubt on the unified scheme in AGNs. Furthermore, the unified models for black hole accretion have weak constraints and well understanding of the ejection process. Especially, in radio-quiet AGNs, corona and wind-like outflows are the two primary radio-emitters except for the jets, while it is unclear how the three processes interplay with each other and which one is in holding the dominance with the accretion flow evolves. Future high-resolution observations are essential to identify the radio origin. Additionally, it may shed light on how jet bases connect with accretion disk and how jet forms. Again, the high-resolution radio observations of intermediate-mass black holes and super-Eddington AGNs are equally important in filling the break between XRBs and AGNs in the fundamental plane of black hole activity and extending it to the super-Eddington regime.

Author details

Xiaolong Yang[1,2]

1 Shanghai Astronomical Observatory, Key Laboratory of Radio Astronomy, Chinese Academy of Sciences, Shanghai, China

2 Shanghai Astronomical Observatory, Shanghai Key Laboratory of Space Navigation and Positioning Techniques, Chinese Academy of Sciences, Shanghai, China

*Address all correspondence to: yangxl@shao.ac.cn

References

[1] Mirabel F. New Astronomy Reviews. 2017;**78**:1. DOI: 10.1016/j.newar.2017.04.002

[2] Kormendy J, Ho LC. Annual Review of Astronomy and Astrophysics. 2013;**51**: 511. DOI: 10.1146/annurev-astro-082708-101811

[3] Fender RP, Belloni TM, Gallo E. Monthly Notices of the Royal Astronomical Society. 2004;**355**:1105. DOI: 10.1111/j.1365-2966.2004.08384.x

[4] Esin AA, McClintock JE, Narayan R. The Astrophysical Journal. 1997;**489**:865. DOI: 10.1086/304829

[5] Ruan JJ, Anderson SF, Eracleous M, Green PJ, Haggard D, MacLeod CL. The Astrophysical Journal. 2019;**883**:76. DOI: 10.3847/1538-4357/ab3c1a

[6] Vierdayanti K, Sadowski A, Mineshige S, Bursa M. Monthly Notices of the Royal Astronomical Society. 2013; **436**:71. DOI: 10.1093/mnras/stt1467

[7] Gladstone JC, Roberts TP, Done C. Monthly Notices of the Royal Astronomical Society. 2009;**397**:1836. DOI: 10.1111/j.1365-2966.2009.15123.x

[8] Müller A. Ph.D. Thesis, 2004.

[9] Gallo E, Fender RP, Pooley GG. Monthly Notices of the Royal Astronomical Society. 2003;**344**:60. DOI: 10.1046/j.1365-8711.2003.06791.x

[10] Narayan R, Yi I. The Astrophysical Journal. 1995;**452**:710. DOI: 10.1086/176343

[11] Shakura NI, Sunyaev RA. Astronomy and Astrophysics. 1973;**24**:337

[12] Merloni A, Heinz S, di Matteo T. Monthly Notices of the Royal Astronomical Society. 2003;**345**:1057. DOI: 10.1046/j.1365-2966.2003.07017.x

[13] Yang X, Yao S, Yang J, Ho LC, An T, Wang R. The Astrophysical Journal. 2020;**904**:200. DOI: 10.3847/1538-4357/abb775

[14] Svoboda J, Guainazzi M, Merloni A. Astronomy and Astrophysics. 2017;**603**: A127. DOI: 10.1051/0004-6361/201630181

[15] Falcke H, Körding E, Markoff S. Astronomy and Astrophysics. 2004;**414**: 895. DOI: 10.1051/0004-6361:20031683

[16] Ho LC. The Astrophysical Journal. 2002;**564**:120. DOI: 10.1086/324399

[17] Sikora M, Stawarz L, Lasota J-P. The Astrophysical Journal. 2007;**658**:815. DOI: 10.1086/511972

[18] Broderick JW, Fender RP. Monthly Notices of the Royal Astronomical Society. 2011;**417**:184. DOI: 10.1111/j.1365-2966.2011.19060.x

[19] McHardy IM, Koerding E, Knigge C, Uttley P, Fender RP. Nature. 2006;**444**: 730. DOI: 10.1038/nature05389

[20] Körding EG, Jester S, Fender R. Monthly Notices of the Royal Astronomical Society. 2006;**372**:1366. DOI: 10.1111/j.1365-2966.2006.10954.x

[21] Saikia P, Körding E, Coppejans DL, Falcke H, Williams D, Baldi RD. Astronomy and Astrophysics. 2018;**616**: A152. DOI: 10.1051/0004-6361/201833233

[22] Gültekin K, King AL, Cackett EM, Nyland K, Miller JM, Di Matteo T. The Astrophysical Journal. 2019;**871**:80. DOI: 10.3847/1538-4357/aaf6b9

[23] O'Dea CP, Saikia DJ. Astronomy and Astrophysics Review. 2021;**29**:3. DOI: 10.1007/s00159-021-00131-w

[24] Stawarz L, Ostorero L, Begelman MC, Moderski R, Kataoka J, Wagner S. The Astrophysical Journal. 2008;**680**: 911. DOI: 10.1086/587781

[25] Fan X-L, Bai J-M. The Astrophysical Journal. 2016;**818**:185. DOI: 10.3847/0004-637X/818/2/185

[26] Wójtowicz A, Stawarz I, Cheung CC, Ostorero L, Kosmaczewski E, Siemiginowska A. The Astrophysical Journal. 2020;**892**:116. DOI: 10.3847/1538-4357/ab7930

[27] Liao M, Gu M, Zhou M, Chen L. Monthly Notices of the Royal Astronomical Society. 2020;**497**:482. DOI: 10.1093/mnras/staa1559

[28] Meier D. The Astrophysical Journal. 1996;**459**:185. DOI: 10.1086/176881

[29] Fender R, Corbel S, Tzioumis T, McIntyre V, Campbell-Wilson D, Nowak M. The Astrophysical Journal. 1999;**519**:L165. DOI: 10.1086/312128

[30] Neilsen J, Lee JC. Nature. 2009;**458**: 481. DOI: 10.1038/nature07680

[31] Sobolewska MA, Siemiginowska A, Gierliski, M. Monthly Notices of the Royal Astronomical Society. 2011;**413**:2259. DOI: 10.1111/j.1365-2966.2011.18302.x

[32] Lusso E, Comastri A, Vignali C, Zamorani G, Brusa M, Gilli R. Astronomy and Astrophysics. 2010;**512**: A34. DOI: 10.1051/0004-6361/200913298

[33] Grupe D, Komossa S, Leighly KM, Page KL. The Astrophysical Journal Supplement Series. 2010;**187**:64. DOI: 10.1088/0067-0049/187/1/64

[34] Vignali C, Brandt WN, Schneider DP. The Astronomical Journal. 2003;**125**: 433. DOI: 10.1086/345973

[35] Strateva IV, Brandt WN, Schneider DP, Vanden Berk DG, Vignali C. The Astronomical Journal. 2005;**130**:387. DOI: 10.1086/431247

[36] Steffen AT, Strateva I, Brandt WN, Alexander DM, Koekemoer AM, Lehmer BD. The Astronomical Journal. 2006;**131**: 2826. DOI: 10.1086/503627

[37] Just DW, Brandt WN, Shemmer O, Steffen AT, Schneider DP, Chartas G. The Astrophysical Journal. 2007;**665**: 1004. DOI: 10.1086/519990

[38] Wu J, Vanden Berk D, Grupe D, Koch S, Gelbord J, Schneider DP. The Astrophysical Journal Supplement Series. 2012;**201**:10. DOI: 10.1088/0067-0049/201/2/10

[39] Trichas M, Green PJ, Constantin A, Aldcroft T, Kalfountzou E, Sobolewska M. The Astrophysical Journal. 2013; **778**:188. DOI: 10.1088/0004-637X/778/2/188

[40] Vagnetti F, Antonucci M, Trevese D. Astronomy and Astrophysics. 2013;**550**: A71. DOI: 10.1051/0004-6361/201220443

[41] Volonteri M, Rees MJ. The Astrophysical Journal. 2005;**633**:624. DOI: 10.1086/466521

[42] Miller-Jones JCA, Tetarenko AJ, Sivakoff GR, Middleton MJ, Altamirano D, Anderson GE. Nature. 2019;**569**:374. DOI: 10.1038/s41586-019-1152-0

[43] van den Eijnden J, Degenaar N, Russell TD, Wijnands R, Miller-Jones JCA, Sivakoff GR. Nature. 2018;**562**:233. DOI: 10.1038/s41586-018-0524-1

[44] Mattila S, Pérez-Torres M, Efstathiou A, Mimica P, Fraser M, Kankare E. Science. 2018;**361**:482. DOI: 10.1126/science.aao4669

[45] Kaaret P, Feng H, Roberts TP. Annual Review of Astronomy and Astrophysics. 2017;**55**:303. DOI: 10.1146/annurev-astro-091916-055259

[46] Zauderer BA, Berger E, Soderberg AM, Loeb A, Narayan R, Frail DA. Nature. 2011;**476**:425. DOI: 10.1038/nature10366

[47] Burrows DN, Kennea JA, Ghisellini G, Mangano V, Zhang B, Page KL. Nature. 2011;**476**:421. DOI: 10.1038/nature10374

[48] Bloom JS, Giannios D, Metzger BD, Cenko SB, Perley DA, Butler NR. Science. 2011;**333**:203. DOI: 10.1126/science.1207150

[49] Levan AJ, Tanvir NR, Cenko SB, Perley DA, Wiersema K, Bloom JS. Science. 2011;**333**:199. DOI: 10.1126/science.1207143

[50] Done C, Wardziński G, Gierliński M. Monthly Notices of the Royal Astronomical Society. 2004;**349**: 393. DOI: 10.1111/j.1365-2966.2004.07545.x

[51] Fabrika S. Astrophysics and Space Physics Reviews. 2004;**12**:1

[52] Blandford R, Meier D, Readhead A. Annual Review of Astronomy and Astrophysics. 2019;**57**:467. DOI: 10.1146/annurev-astro-081817-051948

[53] Hjellming RM, Johnston KJ. Nature. 1981;**290**:100. DOI: 10.1038/290100a0

[54] Mirabel IF, Rodrguez LF. Nature. 1994;**371**:46. DOI: 10.1038/371046a0

[55] Yang J, An T, Zheng F, Baan WA, Paragi Z, Mohan P. Monthly Notices of the Royal Astronomical Society. 2019; **482**:1701. DOI: 10.1093/mnras/sty2798

[56] Yao S, Yang X, Gu M, An T, Yang J, Ho LC. Monthly Notices of the Royal Astronomical Society. 2021;**508**:1305. DOI: 10.1093/mnras/stab2651

[57] Fan L, Chen W, An T, Xie F-G, Han Y, Knudsen KK. The Astrophysical Journal. 2020;**905**:L32. DOI: 10.3847/2041-8213/abcebf

[58] Yang J, Paragi Z, An T, Baan WA, Mohan P, Liu X. Monthly Notices of the Royal Astronomical Society. 2020;**494**: 1744. DOI: 10.1093/mnras/staa836

[59] Laor A, Behar E. Monthly Notices of the Royal Astronomical Society. 2008; **390**:847. DOI: 10.1111/j.1365-2966.2008.13806.x

[60] Zakamska NL, Greene JE. Monthly Notices of the Royal Astronomical Society. 2014;**442**:784. DOI: 10.1093/mnras/stu842

[61] Nims J, Quataert E, Faucher-Giguère C-A. Monthly Notices of the Royal Astronomical Society. 2015;**447**:3612. DOI: 10.1093/mnras/stu2648

[62] Panessa F, Baldi RD, Laor A, Padovani P, Behar E, McHardy I. Nature Astronomy. 2019;**3**:387. DOI: 10.1038/s41550-019-0765-4

[63] Paragi Z, Vermeulen RC, Fejes I, Schilizzi RT, Spencer RE, Stirling AM. Astronomy and Astrophysics. 1999;**348**: 910

[64] Blundell KM, Mioduszewski AJ, Muxlow TWB, Podsiadlowski P, Rupen MP. The Astrophysical Journal. 2001; **562**:L79. DOI: 10.1086/324573

[65] Jeffrey RM, Blundell KM, Trushkin SA, Mioduszewski AJ. Monthly Notices of the Royal Astronomical Society. 2016;**461**:312. DOI: 10.1093/mnras/stw1322

[66] Greene JE, Strader J, Ho LC. Annual Review of Astronomy and Astrophysics. 2020;**58**:257. DOI: 10.1146/annurev-astro-032620-021835

[67] Wu X-B, Wang F, Fan X, Yi W, Zuo W, Bian F. Nature. 2015;**518**:512. DOI: 10.1038/nature14241

[68] Bañados E, Venemans BP, Mazzucchelli C, Farina EP, Walter F, Wang F. Nature. 2018;**553**:473. DOI: 10.1038/nature25180

[69] Woo J-H, Cho H, Gallo E, Hodges-Kluck E, Le HAN, Shin J. Nature Astronomy. 2019;**3**:755. DOI: 10.1038/s41550-019-0790-3

[70] Davis TA, Nguyen DD, Seth AC, Greene JE, Nyland K, Barth AJ. Monthly Notices of the Royal Astronomical Society. 2020;**496**:4061. DOI: 10.1093/mnras/staa1567

[71] Nyland K, Marvil J, Wrobel JM, Young LM, Zauderer BA. The Astrophysical Journal. 2012;**753**:103. DOI: 10.1088/0004-637X/753/2/103

[72] Paragi Z, Frey S, Kaaret P, Cseh D, Overzier R, Kharb P. The Astrophysical Journal. 2014;**791**:2. DOI: 10.1088/0004-637X/791/1/2

[73] Narayan R, Yi I. The Astrophysical Journal. 1994;**428**:L13. DOI: 10.1086/187381

[74] Czerny B, Rózanska A, Kuraszkiewicz J. Astronomy and Astrophysics. 2004;**428**:39. DOI: 10.1051/0004-6361:20040487

[75] Narayan R, Yi I. The Astrophysical Journal. 1995;**444**:231. DOI: 10.1086/175599

[76] Narayan R. Astrophysics and Space Science. 2005;**300**:177. DOI: 10.1007/s10509-005-1178-7

[77] Wu Q, Cao X. The Astrophysical Journal. 2005;**621**:130. DOI: 10.1086/427428

[78] Nyland K, Davis TA, Nguyen DD, Seth A, Wrobel JM, Kamble A. The Astrophysical Journal. 2017;**845**:50. DOI: 10.3847/1538-4357/aa7ecf

[79] Molina M, Reines AE, Greene JE, Darling J, Condon JJ. The Astrophysical Journal. 2021;**910**:5. DOI: 10.3847/1538-4357/abe120

[80] Fischer TC, Secrest NJ, Johnson MC, Dorland BN, Cigan PJ, Fernandez LC. The Astrophysical Journal. 2021;**906**:88. DOI: 10.3847/1538-4357/abca3c

[81] Ho LC, Terashima Y, Ulvestad JS. The Astrophysical Journal. 2003;**589**:783. DOI: 10.1086/374738

[82] Yang Y, Li Z, Sjouwerman LO, Wang QD, Gu Q, Kraft RP. The Astrophysical Journal. 2015;**807**:L19. DOI: 10.1088/2041-8205/807/1/L19

[83] Peng S, Li Z, Sjouwerman LO, Yang Y, Xie F, Yuan F. The Astrophysical Journal. 2020;**894**:61. DOI: 10.3847/1538-4357/ab855d

[84] Wrobel JM, Ho LC. The Astrophysical Journal. 2006;**646**:L95. DOI: 10.1086/507102

[85] Yuan F, Narayan R. Annual Review of Astronomy and Astrophysics. 2014;**52**:529. DOI: 10.1146/annurev-astro-082812-141003

[86] LaMassa SM, Cales S, Moran EC, Myers AD, Richards GT, Eracleous M. The Astrophysical Journal. 2015;**800**:144. DOI: 10.1088/0004-637X/800/2/144

[87] Urry CM, Padovani P. Publications of the Astronomical Society of the Pacific. 1995;**107**:803. DOI: 10.1086/133630

[88] Antonucci R. Annual Review of Astronomy and Astrophysics. 1993;**31**:473. DOI: 10.1146/annurev.aa.31.090193.002353

[89] Nicholls B, Brimacombe J, Kiyota S, Stone G, Cruz I, Trappett D. Astronomer's Telegram. 2018;**2018**:11391

[90] Trakhtenbrot B, Arcavi I, MacLeod CL, Ricci C, Kara E, Graham ML. The Astrophysical Journal. 2019;**883**:94. DOI: 10.3847/1538-4357/ab39e4

[91] Kara E, Loewenstein M, Remillard RA, Gendreau K, Arzoumanian Z, Arcavi I. Astronomer's Telegram. 2018;**2018**:12169

[92] Ricci C, Loewenstein M, Kara E, Remillard R, Trakhtenbrot B, Arcavi I. The Astrophysical Journal Supplement Series. 2021;**255**:7. DOI: 10.3847/1538-4365/abe94b

[93] Boller T, Voges W, Dennefeld M, Lehmann I, Predehl P, Burwitz V. Astronomy and Astrophysics. 2003;**397**:557. DOI: 10.1051/0004-6361:20021520

[94] Gallo LC, MacMackin C, Vasudevan R, Cackett EM, Fabian AC, Panessa F. Monthly Notices of the Royal Astronomical Society. 2013;**433**:421. DOI: 10.1093/mnras/stt735

[95] Ricci C, Kara E, Loewenstein M, Trakhtenbrot B, Arcavi I, Remillard R. The Astrophysical Journal. 2020;**898**:L1. DOI: 10.3847/2041-8213/ab91a1

[96] Scepi N, Begelman MC, Dexter J. Monthly Notices of the Royal Astronomical Society. 2021;**502**:L50. DOI: 10.1093/mnrasl/slab002

[97] Spada M, Ghisellini G, Lazzati D, Celotti A. Monthly Notices of the Royal Astronomical Society. 2001;**325**:1559. DOI: 10.1046/j.1365-8711.2001.04557.x

[98] Mohan P, An T, Zhang Y, Yang J, Yang X, Wang A. The Astrophysical Journal. 2022;**927**:74. DOI: 10.3847/1538-4357/ac4cb2

[99] Yang X, et al. In preparation

Chapter 4

Wave Propagation Theory Denies the Big Bang

Peter Y.P. Chen

Abstract

Problems related to Big Bang because of the Doppler interpretation of cosmological redshift have not been resolved up to recent years. The "tired light" theory proposes an energy loss model that has its own limitations. Chen in 2020 and 2021 proposed to treat light propagation through the space just as a field problem involving electromagnetic waves and governed by the well-known nonlinear Schrödinger (NLS) equation. The space is not a vacuum and is sparsely populated with matters. Electromagnetic waves traveling through the space will undergo changes as predicted by the NLS equation involving a linear dispersion and a nonlinear self-phase focusing terms. Using the cosmological principle, the coefficients associated with these terms could be constants but extremely small in value. Special numerical methods have been developed and could be used to find both bright and dark soliton-like solutions for the NLS equation that are stable and could travel through the extremely long distance involved. These solutions clearly show the redshift is linearly proportional to distance traveled for both bright and dark solitons. The conclusion is that redshift (and blue shift) is an innate nature of light traveling through the space.

Keywords: wave propagation, nonlinear Schrödinger equation, tired light, big bang, cosmological redshift, Hubble's law

1. Introduction

Using electromagnetic wave propagation theory, light transmitting through space is a field problem governed by the well-known nonlinear Schrödinger (NLS) equation. Transmission characteristics, such as redshift, can be deduced from the solution of this NLS equation instead of space expansion, like in the Big Bang, or by energy losses, such as in "tired light" theory. Until recently, the NLS equation has not been solved under conditions appropriate for space. The NLS equation involves two system parameters: a constant coefficient for the linear dispersion term and another constant for the nonlinear self-phase focusing term. As space is sparsely populated with matter, both these constants are extremely small in value. On the other hand, light transitions through space could involve distances of thousands of light years. Chen in 2020 and 2022 has overcome these numerical difficulties in dealing with extremely small and large numbers and developed a numerical method that provides stable soliton solutions that have particle-like characteristics. Solitons can survive the long journey to reach us notwithstanding what could be encountered on the way. After reviewing those numerical solutions, it is clear that any wavelength changes in the propagation of light are linearly proportional to distance traveled. This fact has been observed by astronomers for many years. However, the change in wavelength is solely due to the innate nature of propagation of electromagnetic waves in a medium.

2. Big bang and "tired light"

From the historical beginning, misgivings about the hypothesis space expansion have produced the Doppler effect which causes redshifts in starlight and have been summed up by Shoa [1]. This hypothesis together with observed redshifts leads to Hubble's law [2], which provides the main scientific evidence for the Big Bang theory. Since that time, no one has questioned this fundamental hypothesis except a small number of opponents, such as proponents of "tired light" [1, 3, 4]. During Hubble's time, only redshifts smaller than 0.1 were involved; then, this hypothesis would not invoke too great a controversy. After all, astronomers find that the Hubble's law is a useful empirical relation that could be used to help them to manage many of their astronomical observations. With small redshifts, both Special and General Relativity Theories give similar predictions [5]. Yet today, we are dealing with redshifts larger than 3 and up to 10 or higher. The new problem is that both relativity theories predict for cases involving far distances with much higher values for redshift than that found in the Hubble's law [5]. There are also problems in physical interpretations associated with large redshifts. For example, for redshift greater than 1, the light emitting source would have to travel at speed greater than light. For an object traveling with such a speed, we would lose sight of it, leading to the assumption there is an event horizon beyond that no object can be seen. But in reality, we are still seeing objects having redshift much greater than 1, and recently NASA has sent out a space telescope, Webb, with specific objectives to observe those distant objects.

There is also a problem with how a universe could support billions of multi-solar mass objects all accelerating outward at speed much greater than light. To maintain such a system requires massive amount of energy. It is creative for someone to suggest there is some unknown energy called "dark energy." But to support this preposition, the universe must consist of some 67% of this dark energy with all the visible masses and all forms of known energy making up only some 5%. There is already a problem in being able to explain how the visible universe came to exist; it would be much more difficult to explain how this many times larger dark energy could come to be.

It should be noted that, as recently as 2020, astronomers over the world working on redshift have called for new physics to explain this phenomenon [6].

Tired light theory uses a different hypothesis that redshift is due to energy loss because of interaction between photons in light waves and material particles present in space, such as hydrogen. Although it is claimed that such a hypothesis is based on physical principles that consists of (a) electromagnetic field theory, (b) the mass-energy equivalence, (c) the quantum light theory, and (d) the Lorentz theory [1], it is difficult to see how those principles have been applied to give the final expression for redshift in the tired light theory. Because matter-energy equivalence is an accepted physical principle, it cannot be used to justify the statement "the electromagnetic field and material particles can be considered the same thing," as reported [1]. Similarly, assertive statements, like "The average wavelength of the visible light is 5.5×10^{-7} m, being the diameter of a photon" [1], is difficult to justify as photon is an arbitrarily chosen unit associated with the energy, not the wavelength of a light wave. As dark spectral lines are also redshifted, it is difficult to justify that a dark pulse could suffer energy loss. There is arbitrariness in deriving some of the mathematical statements as well.

Without any doubt, a comprehensive electromagnetic field theory for principle (a) will cover the rest of the physical principles, (b) to (d), as stipulated in the tired light theory [1].

3. Field equation for electromagnetics wave propagation: the NLS equation

The well-known NLS equation involves u, the slowly varying envelope of the axial electric field,

$$u_x - \frac{i}{2}D(x)u_{tt} - i\gamma|u|^2u = 0 \tag{1}$$

where $D(x), \gamma, x$, and t are the dispersion coefficient, self-phase modulation parameters, the spatial propagation distance, and temporal local time, respectively. Using scaling factors, x_o and t_o, so that the solution may be generally applicable to various physical systems:

$$x^* = \frac{x}{x_o}, t^* = \frac{t}{t_o} \tag{2}$$

Then, with

$$D^* = \frac{Dx_o}{t_o^2}, u^* = (\gamma x_o)^{0.5} u \tag{3}$$

Eq. (1) becomes dimensionless with $\gamma = 1$ and the superscript * omitted.

3.1 Numerical solution method

The numerical approach is based on the reduction of Eq. (1) into a set of simultaneous first-order ordinary differential equations (ODEs) by the Lanczos-Chebyshev pseudospectral (LCP) method [7, 8], and the set of simultaneous ordinary differential equations (ODEs) is solved by a stable forward marching procedure. The temporal local time, t, is mapped into a numerical window $[-1, 1]$. The solution is written as an economized power series [7, 8]:

$$u(t,x) = \sum_{n=0}^{N} u_n(x)t^n \tag{4}$$

The derivatives can be obtained from Eq. (4) by means of the term-by-term differentiation to give

$$u'(t,x) = \sum_{n=1}^{N} nu_n(x)t^{n-1}, u''(t,x) = \sum_{n=2}^{N} n(n-1)u_n(x)t^{n-2}. \tag{5}$$

Chose $N - 1$ collocation points, x_i, within the interval $[-1,1]$, that are the roots of the Chebyshev polynomial $T_{N-2}(x)$ [7, 8]:

$$t_n = -\cos\left(\frac{(2n+1)\pi}{2(N-2)}\right), n = 0, \dots, N-2 \tag{6}$$

Together with the two boundary conditions, a set of $N + 1$ ODEs is obtained.

As u is a spike, to give the needed accuracy, the computational domain in the t-direction needs to be divided into K divisions. Use a N-order power series for each subdivision, the set of ODE is in the form:

$$AU_x(x) - iLU(x) = iQ(x, U) \tag{7}$$

where $\boldsymbol{U}$ is a $[(N + 1) \times K]$ vector consisting of the coefficients of the power series used. For numerical integration in the x-direction, we have used the unconditionally stable and implicit equations. With step size Δx in the propagation distance,

$$A\left(U^{m+1} + U^m\right) - \frac{i\Delta x}{2}\left[L\left(U^{m+1} + U^m\right)\right] = \frac{i\Delta x}{2}\left[Q\left(x, U^{m+1}\right) + Q(x, U^m)\right] \tag{8}$$

where U^m is the value of U at step m. Because of the term $\boldsymbol{Q}\left(x, U^{m+1}\right)$ in the right-hand side (RHS), Eq. (8) is nonlinear; it must be solved by an iterative procedure. Since both operators, $\boldsymbol{A}$ and $\boldsymbol{L}$, at the LHS are linear, for the entire case history, the matrix inversion needs not be done at every step.

3.2 The exactly periodic (EP) soliton solution

The NLS equation is a robust system that provides countless solutions depending on the many variants, the system parameters, and boundary conditions used [9, 10]. As an initial value problem, the initial input also occupies a vital role. For solutions that may be classified as solitons, they are stationary waves that oscillate and repeat themselves over a soliton period along the propagation distance. However, the period could be controlled by specially designed periodic system parameters. One widely used design is the dispersion managed (DM) systems, where a period consists of two halves in length and each with a dispersion coefficient of opposite sign. Making use of this characteristic, the periodic solution could be found by the shooting method that is an iterative algorithm using for a dispersion map [9, 10]:

$$u_{in}^{i+1} = 0.5\left(u_{in}^i + u_{out}^i\right) \tag{9}$$

where u_{in}, and u_{out} are the input and output pulse to the dispersion map, respectively, and the superscript i denote the iteration number.

The DM solitons are used in the design of long-distance optical transmission systems. They are used in this chapter to find out the propagation characteristic in each half of the dispersion map.

3.3 Bright soliton solution

It is necessary to have the initial input pulse close to a bright EP soliton [9]. By using trial and error, a Gaussian pulse [9] is chosen:

$$u(t,0) = \beta \exp\left[-\alpha(t-0.5L)^2\right] \tag{10}$$

where L is the given length for t, α is an arbitrarily chosen constant, and β is an adjusting parameter to give a specified pulse energy, E:

$$E(x) = \int_{-\frac{L}{2}}^{\frac{L}{2}} \left(|u(t,x)|^2\right) dt \tag{11}$$

It is important to set the boundary conditions as:

$$u(t,x) = 1000\frac{\partial u}{\partial t} - u(t,x), \; at\; x = \pm 0.5L \tag{12}$$

The large constant associated with the derivative term will force u to assume a near zero value with zero gradient so that the reflection at the boundaries is eliminated.

An example used L = 40, K = 10, N = 20, Δx = 0.001, D = 0.1, α = 1.5, and E = 0.25. The dispersion map has a length of 6. How the solutions converged to periodic and linear wavelength changes can be seen in the plots in **Figure 1**. The distance, x, shown in this plot is the cumulated distance with each iteration, and the pulse traveled through a distance equal to the dispersion map length of 6 (or 12). As the step size is 0.0005, each iteration generates 12,000 pulse histories, but only every 40th history is shown in **Figure 1**.

The pulse histories show evidence of convergence. Moreover, when doubling the period length from 6.0 to 12.0, the pulse has shown the same linear broadening and narrowing characteristics. **Figure 2** shows the changing pulse shapes in traveling through a period.

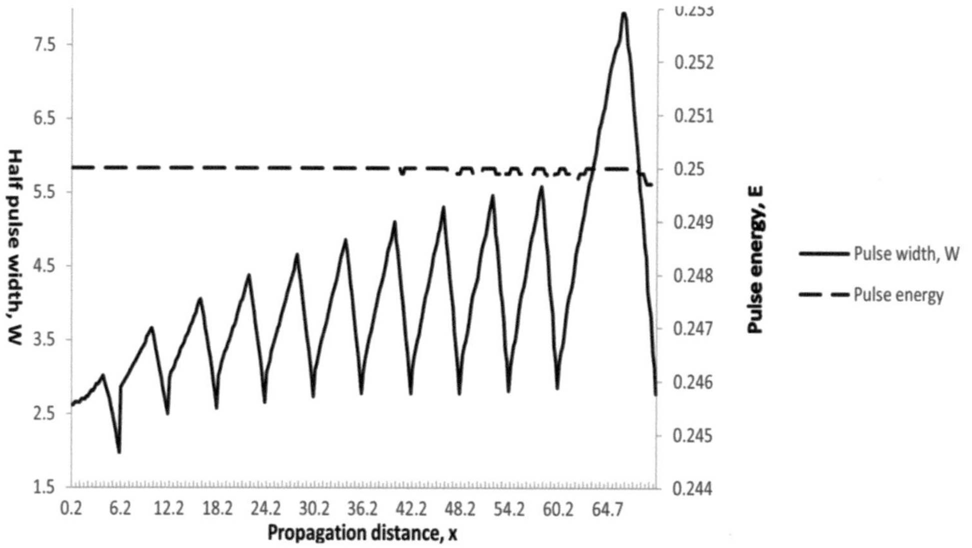

Figure 1.
The convergence of the iterative method.

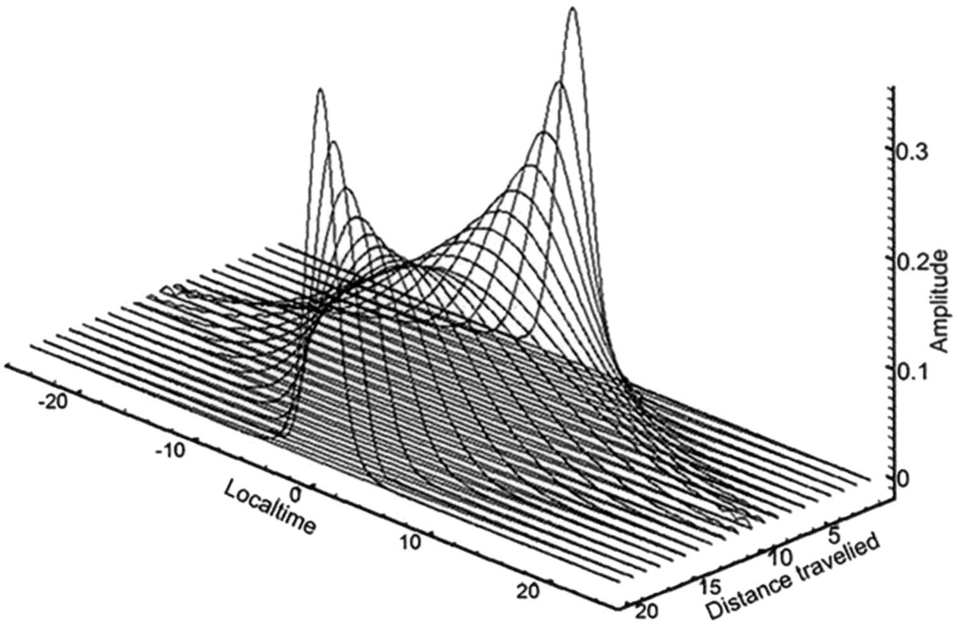

Figure 2.
Broadening and narrowing of pulse width through a dispersion map.

3.4 Dark soliton solution

A dark soliton [10] is obtained if the initial input pulse is taken to be

$$u(t,0) = \beta\ \exp\left(1 - \left[\alpha(t - 0.5L)^2\right]\right) \tag{13}$$

with boundary conditions:

$$u(t, \pm 0.5L) = 0 \tag{14}$$

In an example [10], the followings are used: D = 0.4, L = 40, E = 2.0, and α = 0.15. The constant, β, is found from the pulse energy. The dispersion map is 6.0 in length with a positive D for the first half and a negative one for the second half. The step size along with the propagation distance, Δx = 0.0005. To cater for the special shape of a soliton and the fact pulse width is changing, the size of the numerical window must be carefully chosen. The observed propagating characteristics of the pulse width are expanding in the first half of the dispersion map, where dispersion is positive and contracting in the second half where dispersion is negative. **Figure 3** shows how the iteration has converge to an EP solution even when the period is increased to twice of its length in the last cycle. Although after 40 iteration cycle, there was still an approximately 0.1% decrease in the pulse energy per cycle; however, the pulse histories show clear evidence of convergence (**Figure 3**). **Figure 4** shows the change of pulse shape in the first half of the dispersion map.

3.5 Calibration to redshift-distance relationship

In astronomy, redshift, z, is defined by wavelength changes:

DOI: http://dx.doi.org/10.5772/intechopen.103848

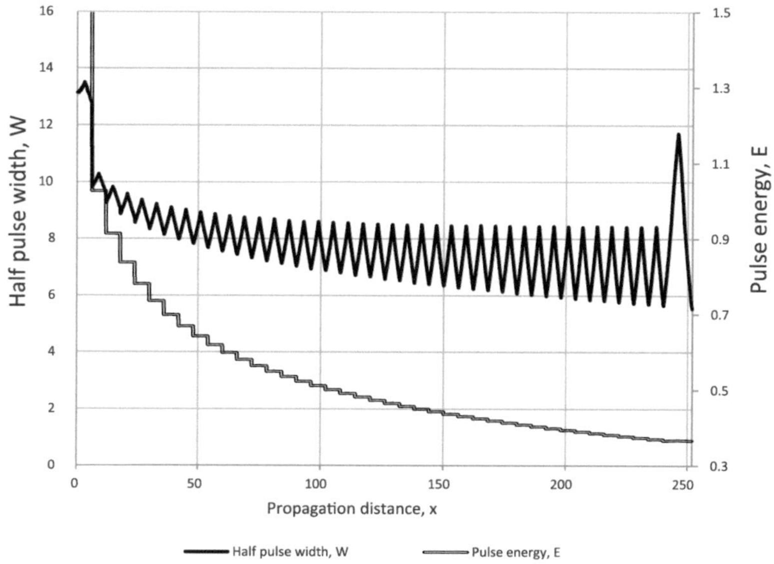

Figure 3.
Iteration leading to periodic dark soliton solution.

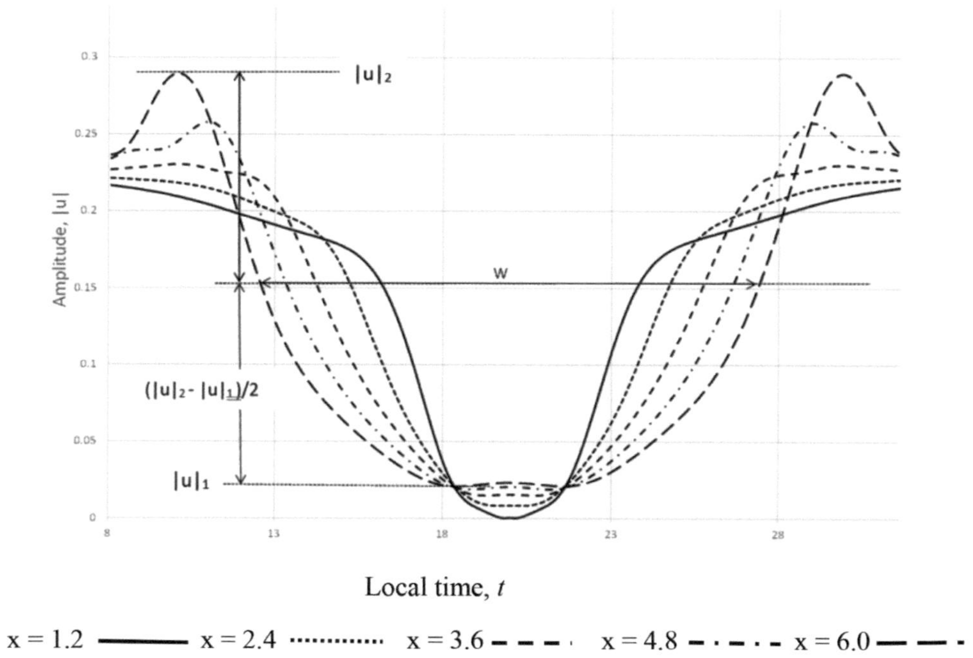

Figure 4.
Transmission of a dark EP soliton in the normal dispersion segment.

$$z = \frac{\lambda_2 - \lambda_1}{\lambda_1} \tag{15}$$

where λ_1 and λ_2 are the starting and ending wavelength of a spectral line. The redshift-distance relationship [10] is known as the Hubble's law:

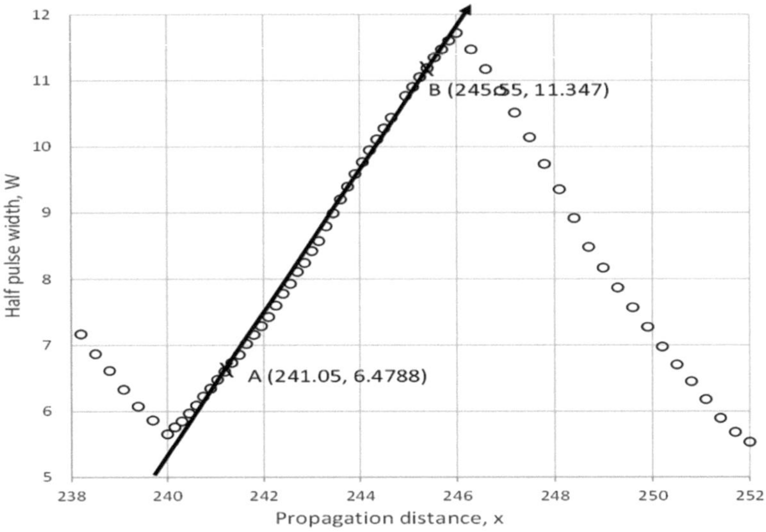

Figure 5.
Linear relationship between W *and* x.

$$z = H_o d/c \tag{16}$$

where H_o is the Hubble's constant and is determined experimentally. Instead of x, the distance d is in unit of Mpc, while c, the velocity of light, is in km/s. As the linear relationship found numerically in Sections 3.2 and 3.3 is in different units, calibration must used to convert the findings to the same as in Eq. (16).

If λ_1 and λ_2 are each proportional to full width at half maximum (FWHM) W_1 and W_2, then the redshift:

$$\text{z} = \frac{\lambda_2 - \lambda_1}{\lambda_1} = \frac{W_2 - W_1}{W_1} \tag{17}$$

Using a larger scale, the last iterative cycle shown in **Figure 3** is replotted to give **Figure** 5. Based on two selected points, A(241.05, 6.4788) and B(245.55, 11.347), the linear relation is found to be

$$W = -254.34 + 1.082x^* \tag{18}$$

As both Eq. (16) and (18) are linear, we could use A and B as calibration points and covert Eq. (18) to the same form as Eq. (16). If point A (x_A = 241.05, W_A = 64,788) is selected as the reference point, using Eq. (17), redshift z_A for reaching the point B (x_B = 245.55, W_B = 11.347) is

$$z_A = \frac{W_B - W_A}{W_A} = 0.7514 \tag{19}$$

With this amount of redshift and using Eq. (16) with H_o = 70.0 km/s/Mpc, an unit often used in astronomy, the distance traveled by the light wave would be

$$(d/c)_A = \frac{z_A}{H_o} = 0.01073 \text{ Mpc/c} \tag{20}$$

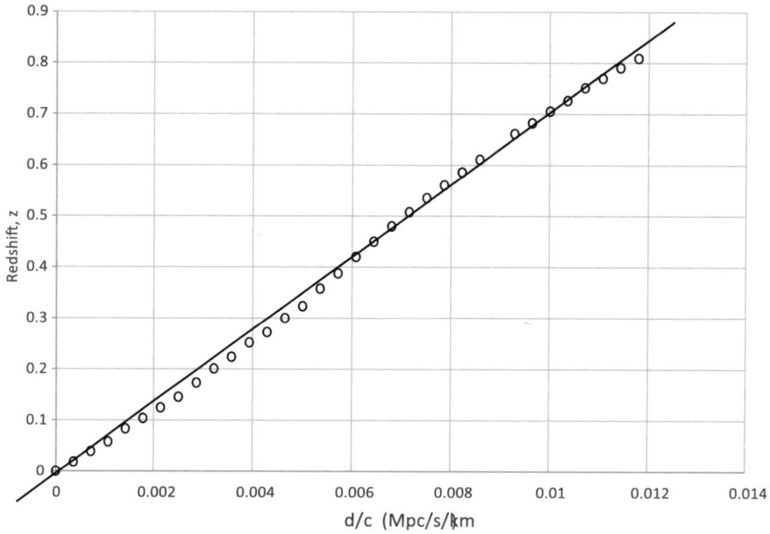

Figure 6.
Redshift, z, versus distance from the earth.

The factor to convert x into d/c:

$$f_x = \frac{(d/c)_A}{(x_B - X_{A)}} = 0.002384 \text{ Mpc/c} \tag{21}$$

Applying conversions to the data points over the pulse width expanding segment AB in **Figure 5**, the new calibrated plot, **Figure 6**, confirms that our results have the same linear relationship as given by the Hubble's Law. It should be noted that as numerical solutions contain inaccuracies and noises, some data points are not exactly on the linear line.

To apply, as an example, the numerical simulation to a real physical system [10], that is the space, the dark spectral line due to Lyman-alpha hydrogen has a wavelength of 121.6 nm. The corresponding period is 405 ps. For the chosen reference point, A, W_A = 6.4788 and the temporal time scaling factor:

$$t_o = \frac{405}{2 \times W_A} = 31.25 \text{ ps} \tag{22}$$

For the distance scaling factor, x_o, using Eq. (20):

$$cf_x(x_B - x_R) = d_R \tag{23}$$

where d_R is the physical distance measured from the point corresponding to x_R, and,

$$d_R = (x_B - x_R)x_o,$$

If x_A is the reference point, $x_o = cf_x$ = 0.002384 Mpc, or 9.30×10^{16} km (which is about 10 light years). Now, an estimation of the dispersion coefficient:

$$D = \frac{t_o^2 D^*}{x_o} = 2.10 \times 10^{-15} \text{ps}^2/\text{km} \tag{24}$$

There is insufficient information available to work out the self-phase modulation parameter, γ. If the light source is the same as our sun, the power of light emitted is known to be 3.9×10^{26} W. Assuming, just as an order of study, (1) absorption to form the dark soliton is taking place at a distance of 1000 km away from the surface and the law of one over distance square law is used, and (2) the most of the power is in the short and utter-short spectrum and only 10^{-10} of the power is associated with the hydrogen absorption spectrum, $|u|^2 = 3.9 \times 10^4$ W. Then,

$$\gamma = \frac{|u^*|^2}{x_o|u|^2} = 1.9 \times 10^{-26}/\text{mW} \tag{25}$$

4. Remarks

Based on the cosmological principle [5] that the universe is both homogeneous and isotropic, it is justifiable to use constant system parameters for the NLS equation, especially in dealing with solitons, where pulse energy is confined to a narrow spectral width. It is well known in physics, for example, that the dispersion coefficient varies with wavelength. But, when solitons are used in optical communication, a single group velocity dispersion coefficient is always used. It should be noted that the cosmological principle does not deny the existence of local deviations from the averages. Therefore, the propagation theory as described here will predict the averaged redshift while local conditions, such as peculiar velocity, gravity, or concentration, will produce deviations in observed data, for example, redshift. Although the deviations could be positive or negative, they may not cancel out each other due to the cosmic scale involved. Deviations from the Hubble's law, which can sometimes be quite large [6], have been observed by cosmologists in the redshift-distance relationship.

Proponents of the Big Bang theory have considered the cosmic microwave background radiation (CMBR) as the second major piece of scientific evidence. The argument is that the short wavelengths gamma rays at the beginning of the Big Bang had been stretched, due to space expansion, to microwaves. All those CMBR have been trapped in the cosmos ever since that time. With the propagation theory, such an explanation is not needed. Those CMBR are simply electromagnetic waves from far away sources that have been broadened through the distance traveled.

Propagation theory is not intended to explain the origin of the universe. It is simply meant to illustrate that the linear relationship existed between the change of wavelength and distance traveled as found in the numerical solutions of the NLS equation. In addition, just like any other measuring instrument, calibration could be used to change the readings to a particular set of units. There is no new principle involved in this approach.

With numerical methods, the LCP method has been used because of fewer equations are solved. Any other numerical method could be used providing that the system is a fixed length dispersion managed map together with an iteration loop based on Eq. (9).

It should be noted that both the solutions for dark and bright solitons could be used to give the calibrated redshift-distance relationship and **Figure 6** is applicable to z values outside the plot.

5. Conclusion

The NLS equation is a well-used theory for electromagnetic wave propagation. Using its EP soliton solutions, the linear wavelength change versus distance relationship can be established. The solutions could be calibrated to fit the empirically observed Hubble's law. The space between a source and an observer need not be expanding for redshift (or blue shift) to occur as stipulated in the Big Bang theory.

Acknowledgements

The author is grateful to Laura Choong, PhD candidate in the University of New South Wales, Sydney, for reading through the paper and making linguistic corrections and modifications on the chapter.

List of abbreviations

CMBR	Cosmic microwave radiation
DM	Dispersion managedEP Exactly periodic
FWHM	Full width at half maximum
LCP	Lanczos-Chebyshev pseudospectral
LHS	Left-hand side
Mpc	An astrological length unit approximately equals 3.09×10^{19} km
NLS	Nonlinear Schrödinger
RHS	Right-hand side

Author details

Peter Y.P. Chen
Former School of Mechanical and Manufacturing Engineering, University of New South Wales, Sydney, NSW Australia

*Address all correspondence to: peterypchen@yahoo.com.au

References

[1] Shao MH, Wang N, Gao ZF. Tired light denies the big bang. In: Robson BA, editor. Redefining Standard Model Cosmology. London: IntechOpen; 2019. DOI: 10.5772/intechopen.81233

[2] Hubble E. A relation between distance and radial velocity among extra-galactic nebulae. Proceedings of the National Academy of Sciences of the United States of America. 1929;**15**:168-173

[3] Zwicky F. On the red shift of spectral lines through interstellar space. Proceedings of the National Academy of Sciences of the United States of America. 1929;**15**:773-779

[4] Shao M. The energy loss of photons and cosmological redshift. Physics Essays. 2013;**26**(2):183-190

[5] Carroll AO, Ostlie DA. An Introduction to Modern Astrophysics. United Kingdom: Cambridge University Press; 2017. DOI: 10.1017/9781108380980

[6] Valentino ED et al. Cosmology intertwined II: The Hubble constant tension. arXiv. 2020, 2008:11284v4

[7] Chen PYP, Malomed BA. Lanczos–Chebyshev pseudospectral methods for wave-propagation problems. Mathematics and Computers in Simulation. 2012, 2011;**5**:13. DOI: 10.1016/j.matcom.2011.05.013

[8] Chen PYP. The Lanczos-Chebyshev pseudospectral method for solution of differential equations. Applications of Mathematics. 2020;**7**:927-938. DOI: 10.4236/am

[9] Chen PYP. A mathematical model for redshift. Applications of Mathematics. 2020;**11**:146-156. DOI: 10.4236/am.2020.113013

[10] Chen PYP. Propagation of dispersion-managed dark solitons and the novel application to redshift in starlight. Optik. 2022;**251**:168384. DOI: 10.1016/j.ijleo.2021.168384

Section 2

Solar System

Chapter 5

Cryovolcanism in the Solar System and beyond: Considerations on Energy Sources, Geological Aspects, and Astrobiological Perspectives

Georg Hildenbrand, Klaus Paschek, Myriam Schäfer and Michael Hausmann

Abstract

Volcanism based on melting rocks (silicate volcanism) is long known on Earth and has also been found on Jupiter's moon Io. Remnants of this type of volcanism have been identified also on other bodies in the solar system. Energy sources powered by accretion and the decay of radioactive isotopes seem to be dominant mainly inside larger bodies, which have enough volume to accumulate and retain this energy in significant amounts. On the other hand, the impact of tidal forces allows even tiny bodies to melt up and pass into the stage of cryovolcanism. The dependence of tidal heating on the size of the object is minor, but the masses of and the distances to accompanying bodies as well as the inner compositions of the heated body are central factors. Even though Io as an example of a body supporting silicate volcanism is striking, the physics of tidal forces might suggest a relatively high probability for cryovolcanism. This chapter aims at considering the parameters known and objects found so far in our solar system to give insights into where in our system and other planetary systems cryovolcanism might be expected.

Keywords: volcanism, cryovolcanism, tidal forces, radioactivity, low-temperature biotopes, black smoker equivalents

1. Introduction

All types of volcanism known until now are in the solar system. All considerations and models for volcanism in other stellar systems are built upon our knowledge from our own system. New types of volcanism still unthought of, might be a challenging research topic but may not be considered here.

The main aim of this chapter is to consider cryovolcanism powered by tidal heating and its potential in exosystems. As an introduction, for reference and to characterize the main features as a base for better comparison, a rough overview of its counterpart silicate volcanism as well as underlying types of energy sources in the solar system are given.

The most prominent objects in our solar system harboring active volcanoes are both an example of what may be named high-temperature range volcanism as rocks are molten and are apparent in the form of glowing liquid lava on Earth and on the moon Io. This is generally better known under silicious-based/silicate volcanism because silicate is the most dominating component in liquid rocks. The known temperatures rise to about 1600 K on Io in the volcanoes on the surface [1], while on Earth about 1000 K to 1550 K temperature in the lava is reached [2] depending on the composition of the rocks. The temperature of magma below the surface may have still higher temperatures.

These two objects already show us also the main energy categories on which volcanism, as we know it, relies on. For the Earth, it is mainly based on the conserved accretion/contraction energy from its formation, decay of radioactive elements pulled into the mantle and center of the planet by gravity-induced differentiation, and also on friction rising from the resulting tectonic activity [3]. For Io, it is mainly based on tidal heating from the huge tidal forces raised by the gas giant it is orbiting in its crust and upper mantle [3]. Both may gain also energy from friction that is arising from the resulting tectonic activity. Some of the following general considerations may also apply to forms of energy resources. The energy retention behavior (and so also the duration of volcanic activity) is among other factors strongly depending on surface-to-volume ratios regardless of energy source. Bigger objects with lower surface-to-volume ratios are tending to stay hotter for a longer period and are able to sustain volcanism longer. For Earth and Moon during the assumed collision of their precursor bodies Theia and Gaia, a transfer of the core of Theia into the forming core of the Earth may have increased also the amount of heavier and so radioactive elements, increasing the power for volcanism on Earth and by this decreasing it on the Moon. Also during this early phase, tidal heating may have played a much bigger role for both objects, as they have been much closer together [4].

Regarding ancient evolution steps in the solar system, it is important stressing that even much tinier impacts than Theia with Gaia were much more common and have played a stronger role in melting parts of a planet, asteroid, or moon, especially during the late heavy bombardment (LHB). As it may be perceived as an external energy source and is now of little relevance, volcanism by bombardment will not be discussed further.

Considering the long-term evolution of heating sources also leads us to inactive silicate volcanism as the bodies considered are too tiny to have been able to sustain volcanism until now, as on the Moon, Mercury [5], Venus, and Mars. They are covered with lava plains and show also volcanoes, for example, the highest of the solar system, Olympus Mons on Mars. Still, for all these objects, signs for stronger or lesser still ongoing or very recent volcanic/tectonic activity have been found or are discussed (Moon: [6–10]; Mercury: [11–13]; Venus: [14–17]; Mars: [18, 19]). On Mars also a connection to a known type of lower temperature volcanism may already be found as the melting of ice and/or its remnants under a volcano may have been found as well [20–23].

In the case of Venus, a relatively young surface [24] and its own type of tectonics [25] may also indicate a presence of modifying influences on silicate volcanism that are not well known until now. If the missing of water or other solvents (on Venus probably mostly after entering into a runaway greenhouse effect) is a cause for a changed plate tectonic and so volcanism [26–28], also availability and abundance of

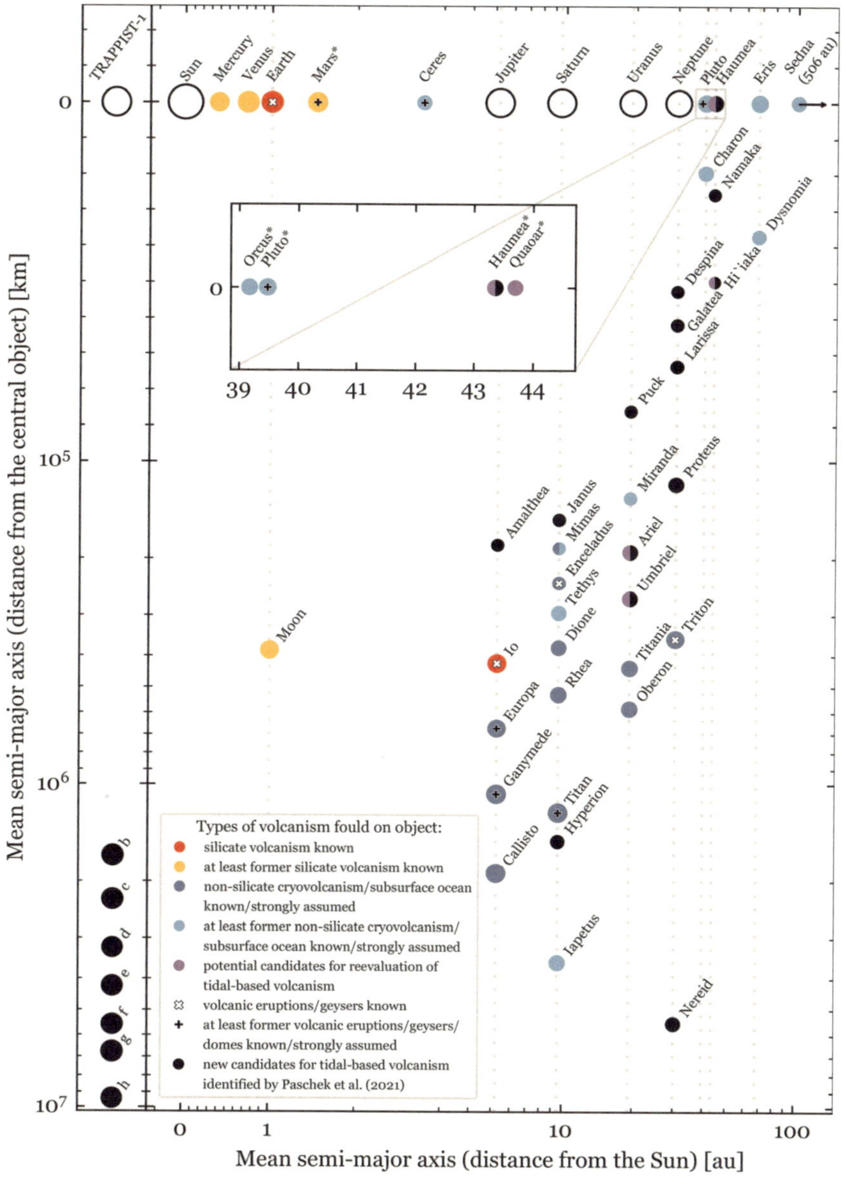

Figure 1.
Overview of types of volcanism identified or assumed on celestial bodies in the solar system (right panel) and the extrasolar planetary system TRAPPIST-1 (left panel). The horizontal axis corresponds to the mean semi-major axis of the orbits as distance to the sun (right panel) and the vertical axis corresponds to the mean semi-major axis of the orbits of moons or exoplanets as the distance from their central object (hosting planet or star TRAPPIST-1) [37–43]. The radii of the circles depicting each object are scaled logarithmically to the actual radii of the celestial bodies [37–39, 41–53]. Please note that close to the dwarf planets Pluto and Haumea the two further dwarf planets Orcus and Quaoar exist. These are shown in the zoomed-in inset panel. Charon is a moon of Pluto, and Namaka and Hi'iaka are moons of Haumea. The dwarf planet Sedna on the right side of the right panel orbits the sun at a mean distance of 506 au, so further away as it is shown here, which is indicated with an arrow. Color-coded is the types of volcanism known or assumed on the objects. ***Table 1*** *Provides an overview of the respective references. The hollow cross marks objects in the solar system for which ongoing volcanic eruptions or geysers are known. The filled plus marks objects in the solar system for which at least former volcanic eruptions, geysers, or domes (remnants of extinct volcanic activity) are known or strongly assumed. Please note that only on Pluto at least former volcanic eruptions are assumed, but not on Orcus. An asterisk (*) at the end of an object's name marks when one or several moons are present but not shown here.*

water, NH_3 or CH_4 have to be considered for modifying silicate volcanism, showing again a link to material and substances beyond rock.

Also, a discussed inhomogeneous distribution of radionuclides as a cause for volcanic activities, for example, on the Moon [6] further highlights a need for deep consideration of how volcanism may be sustained and be modified in behavior.

Moving on outward in the solar system brings us into ranges of asteroids, all of them being tinier than the aforementioned planets and so obviously have cooled and are not maintaining volcanism now. Accretion and radioactive energy seem to be nowadays not important for any type of volcanism in the asteroids. Still, ancient traces of volcanism may be found. The importance of meteorite impacts for melting gets relatively bigger on tinier objects. But also a differing composition of radioactive elements seems to play a bigger role as ^{26}Al [29] and ^{60}Fe [30] seem to have molten these tiny objects and given rise to silicate volcanism. This peroid has made a huge influence on these objects, even though this period may not have been very long, regarding the relatively short half-life of these isotopes.

Entering the realm of the gas giants opens new perspectives. The rocky objects that can show volcanism are now mainly moons, tinier in size but are orbiting much larger gas giants or maybe very close double systems orbiting each other, for example, some TNOs. These conditions open the possibility for tidal heating as the main energy source for volcanoes. Accretion and radioactive energy seem to be nowadays of lesser importance for any type of volcanism in the asteroids, gas giant moons, and beyond in the solar system.

Considering Io as an exception in this range, as we also will show, we encounter two other known examples of volcanism around gas giants that are based on tidal heating, but are now in the lower temperature ranges of cryovolcanism. The moons Enceladus and also Triton have been identified as cryovolcanic worlds [31, 32]. Others show signs of active geology and tectonics, for example, on Europa [33] or Ganymede [34], and are believed to have liquid layers or even oceans of solvents, such as water or NH_3, in their depths and even deeper a basic silicate volcanism.

Regarding this, it becomes easily obvious that a real stable definition of cryovolcanism is not as easy. The aim is mostly trying to focus on volatiles, for example, molten water or methane are thrown out on the surface in an environment colder than their own melting temperature, also even if in greater depths rocks might be quite hot. Earth itself is mostly not being considered as a planet harboring cryovolcanism, even though any volcano under ice known (Iceland) or assumed (Antarctica) and geysers all over the world would fulfill such definitions in winter. Also, mud volcanism (also called "cold" volcanism) being based on mud diapirs and being generally associated with (silicate) volcanism [35, 36], is normally not considered under cryovolcanism.

All these ambiguities in defining cryovolcanism may result from a bias in detecting cryovolcanism on foreign worlds in astronomy or astrophysics. Big eruptions are much easier to observe by optic sensors (on or close to Earth or even on probes) as well as by mass analyzing probes in the proximity of these objects than by constant release of volatiles by tectonics of slowly moving ice shields covering deeper-lying liquids or even silicate volcanism. Also, old remnant structures of previous volcanism may still cover deeper active processes, which is much more problematic to investigate. If we improve our detection capabilities, also our definitions will evolve. Regarding detection and research on cryovolcanic worlds, this all illustrates the strong necessity of modeling based on easier accessible observations, either to understand where we might find such objects with cryovolcanism or what kind of cryovolcanism we might expect. This leads apart from the known active volcanic and cryovolcanic

Object	Primary	Type of volcanism, etc.	Reference
Mercury	Sun	At least former silicate volcanism	[12, 13]
Venus	Sun	At least former silicate volcanism	[15–17]
Earth	Sun	Silicate volcanism; Active volcanic eruptions	
Moon	Earth	At least former silicate volcanism	[9, 10]
Mars	Sun	At least former silicate volcanism; At least former volcanic eruptions/domes	[18, 19]
Ceres	Sun	At least former cryovolcanism; At least former volcanic eruptions/geysers/domes	[54–56]
Io	Jupiter	Silicate volcanism; Active volcanic eruptions	[57]
Europa	Jupiter	Cryovolcanism; At least former volcanic eruptions/geysers/domes	[33, 57–60]
Ganymede	Jupiter	Cryovolcanism; At least former volcanic eruptions/geysers/domes	[34, 57, 61, 62]
Callisto	Jupiter	Cryovolcanism	[57]
Mimas	Saturn	Cryovolcanism/At least former cryovolcanism (debated)	[63–65]
Enceladus	Saturn	Cryovolcanism; Active volcanic eruptions/geysers	[31, 66]
Tethys	Saturn	At least former cryovolcanism	[67–69]
Dione	Saturn	Cryovolcanism	[66]
Rhea	Saturn	Cryovolcanism	[70]
Titan	Saturn	Cryovolcanism; At least former volcanic eruptions/geysers/domes	[71–79]
Iapetus	Saturn	At least former cryovolcanism	[70, 80]
Miranda	Uranus	At least former cryovolcanism	[81–84]
Ariel	Uranus	Potential candidate for at least former cryovolcanism	[81, 82]
Umbriel	Uranus	Potential candidate for at least former cryovolcanism	[81]
Titania	Uranus	Cryovolcanism	[70]
Oberon	Uranus	Cryovolcanism	[70]
Triton	Neptune	Cryovolcanism; Active volcanic eruptions/geysers	[32, 70, 85–89]
Pluto	Sun	At least former cryovolcanism; At least former volcanic eruptions/geysers/domes	[70, 90]
Charon	Pluto	At least former cryovolcanism	[91]
Orcus	Sun	At least former cryovolcanism	[70, 92]
Haumea	Sun	Potential candidate for at least former cryovolcanism	[93]
Hi'iaka	Haumea	Potential candidate for at least former cryovolcanism	[93]
Quaoar	Sun	Potential candidate for at least former cryovolcanism	[94]
Eris	Sun	At least former cryovolcanism	[50, 70]
Dysnomia	Eris	At least former cryovolcanism	[50]
Sedna	Sun	At least former cryovolcanism	[70]

Table 1.
*Celestial objects in the solar system on which different types of volcanism are present or strongly assumed. The last column gives the respective references. References given here were also used to categorize the types of volcanism given in **Figure 1**. For each object, its orbited primary and the types of known or strongly assumed volcanism and eruptions (active or extinct) are listed.*

worlds to a huge list of strongly assumed, mainly cryovolcanic, active as well as inactive worlds (see **Figure 1** and **Table 1**).

All these models are strongly based on energy resources and energy transport. Reconsidering some basic parameters in these models may illuminate some specific aspects of cryovolcanic worlds and offers an insight into basic principles to find general concepts for application in far exoplanetary systems.

2. Volcanism present in the solar system and the extrasolar planetary system TRAPPIST-1

The types of active and inactive volcanism in our own neighborhood are various. **Figure 1** gives an overview of the different types of volcanism found or strongly assumed on celestial objects in the solar system. To classify the different types of (cryo-)volcanism found on objects in the solar system, we distinguish between the case when the respective type of volcanism is active right now and verified (e.g., by measurements of space probes) or strongly assumed due to observations, measurements or theoretical models, and the case when signs of at least former volcanic activity were identified. We also include the (at least former) presence of a liquid subsurface ocean as part of cryovolcanism.

The melting up of a subsurface ocean requires a strong energy source. This is either powered from the interior of the body hinting at the presence of silicate volcanism in its core. Another or even simultaneously occurring energy source can be the deformation by tidal forces of nearby objects, which can liquify silicates or ice and heats up potentially present silicate magma and/or a (subsurface) ocean further. This might result in icy objects in cryovolcanic activity, for example, in the form of geysers penetrating through the ice crust of Saturn's ice-moon Enceladus [31]. By cracking up the ice crust a cryo-form of plate tectonics could be initiated, for example, on Jupiter's ice-moon Europa [33].

In addition, we identify several objects that should be considered as potential candidates for re-evaluation of the potential of tidal-based volcanism based on recent studies. For example, the presence of crystalline water ice and/or ammonia ice on the surface hints at the presence of a mechanism that actively redeposits new material, as crystalline water ice and/or ammonia ice is not stable in the long term in these environments due to destruction by energetic particles (see, e.g., [92–94]).

Domes, which are mountains and bulges in the crust of a celestial object, could be remnants of extinct eruptive volcanoes or could be plumes that do/did not penetrate fully through the crust. We see the identification of domes on the surface of a celestial object as an indicator for at least former eruptive volcanic activity.

Moreover, we included the objects resulting from our recent study [95], which we identified as new and (in the case of the solar system) not yet elsewhere considered candidates for tidal-based volcanism.

3. Considerations on energy from tidal heating

For cryovolcanism, an indispensable prerequisite must be an energy source. In principle, energy could be gained from accretion and contraction during the formation of the planetary object. This process is among other parameters depending on the size of the object (with R being the radius of the object, roughly $\sim R^3$). As radioactive material is incorporated along with this process, equivalent considerations may be done here. Higher volume-to-surface ratios ($\sim R$) minimize cooling effects and allow longer stable heating from inside. Rearrangement of material (e.g., impacts as in LHB, Theia-Gaia events, seeding with ^{26}Al) may change the occurrence, intensity, and also duration of volcanic active phases; inhomogeneities in deposition of material may give rise to local volcanism. The starting composition of radioactive material during formation may differ along with, for example, age of the stellar population. These processes will be complete in the very early phase of a stellar system (roughly 0.5 Gy after formation) and any volcanic activity based on this will evolve based on the then built-up conditions for heating and cooling. Models over several Gys imply significant effects for the heating of liquid volatiles in bigger objects of several hundreds of km radius [96].

This "standard" energy production process might not work in smaller objects where other heating sources are required, for instance, tidal heating, a process occurring in planetary systems with masses closely associated and thus impacting each other. The general principles for tidal heating may be considered as based on many more parameters as for accretion/radioactivity. Aspects of volume-to-surface ratios ($\sim R$) are the same as for accretion and radioactivity, many other parameters differ.

The tidal acceleration A acting on an object's surface is

$$A = \frac{GM}{r^2}\left(\frac{1}{\left(1 \pm \frac{R}{r}\right)^2} - 1\right), \tag{1}$$

G as gravitational constant, M as mass of the influencing object, R as radius of the influenced object, and r as distance between the objects. This can be approximated by a Taylor series expansion to

$$A = \mp 2GM\frac{R}{r^3}. \tag{2}$$

Therefore, the tidal force will go with $\sim R$ (for details and elaborated calculations see [95]). The energy transfer and average dissipation rate gets based on even more parameters and may mostly be assumed by $\sim R^5$ [97–103].

$$\dot{E} = -\frac{21}{2}\frac{k_2}{Q}\frac{n^5 R^5}{G^*}e^2, \tag{3}$$

$\dot{E}$ as rate for tidal energy dissipating, G^* as gravitational constant, k_2 as Love number, and Q dissipation function of the satellite. $\frac{k_2}{Q}$ is telling how "effectively" energy is transferred on the satellite and how this leads to heating. Models with k_2 are mainly used, but also models with "higher" Love numbers as k_3, k_4, or k_6 may be considered reasonable for special systems [97, 104–106].

Q is in the range from 10 to 500 are found for the terrestrial planets and satellites of the major planets. On the other hand, Q for the major planets is always larger than $6 \cdot 10^4$ [106].

Trying to figure out further principles for tidal heating we may approach this by considering when tidal heating may really be minimized.

A body that is tidally locked on an orbit with eccentricity $e = 0$ will not have any type of tidal energy dissipating. Locking will occur in ranges of

$$t_{\text{lock}} = \frac{\omega a^6 I Q}{3G^* {m_p}^2 R^5 k_2}, \tag{4}$$

G^*, k_2, Q, R as above, ω as initial spin rate, a for the semi-major axis of the orbit of the satellite around the planet/partner, m_s as mass of the satellite, m_p as mass of the planet/partner, and I as momentum of inertia [107] (see pages 169–170 of this article; Formula (9) is quoted here, which comes from ref. [108]), with $I \approx 0.4 m_s R^2$:

$$t_{\text{lock}} \approx \frac{\omega a^6 0.4 m_s R^2 Q}{3G^* {m_p}^2 R^5 k_2} = \frac{0.4\omega}{3G^*} \frac{Q}{k_2} \frac{a^6 m_s}{{m_p}^2 R^3}. \tag{5}$$

With $m_s = \frac{4\pi}{3} \rho R^3$ and ρ as density of the satellite:

$$t_{\text{lock}} \approx \frac{1.6\pi\omega}{9G^*} \frac{Q}{k_2} \rho \frac{a^6}{{m_p}^2}. \tag{6}$$

Apart from ω resulting from the formation process, $\frac{Q}{k_2}$ and ρ, as parameters for interior composition and "behavior" in heating, m_p and especially a seem to strongly influence the period in which tidal heating may be possible.

The moon Io is actually tidally locked and would be on a far bigger orbit with eccentricity $e = 0$ and so no volcanism at all would occur, if its accompanying moons would not have been influencing it and are distracting it from a round orbit [109, 110].

But a may also change over longer periods "on its own" and may so become important regarding the period for tidal heating and so volcanism. This results from an effect of energy transfer by tidal forces beyond heating, yielding a change of orbital velocity because of tidal acceleration or tidal deceleration.

As the energy transfer resulting in heating is not the only effect, tidal acceleration and also tidal deceleration may occur and by changes in velocity, change the orbit of the objects. For tidal acceleration this will bring objects to farther orbits, moving them out of the possible zone for tidal heating, for tidal deceleration, this will lower the orbits and so either crushing the objects when crossing the Roche limit or crashing them on the body which they are orbiting, as it is assumed for Triton [111–113]. These effects have also an impact via changes in the semi-major axis a on t_{lock} . The changes by tidal acceleration/deceleration are still tiny in our system and so changes in t_{lock} maybe on larger scales [111–113].

All these aspects make it obvious how variable volcanism based on tidal heating may be. The discovery of so powered cryovolcanism on the moons Enceladus and also Triton has been quite surprising and many proofs or hints for active or inactive volcanism, of any kind, may have still not been found in the region of the asteroid belt and beyond. A general overview of both silicate volcanism and cryovolcanism is given in **Figure 2**. All sketches of phases given may be powered by both accretion and radioactivity or by tidal heating. Especially if objects are big or young enough, we may also consider overlap of both power types. Known objects in our own system cover only some of these sketches, but still, we do not have proof of volcanism on all objects being considered and, as discussed, some may be cryovolcanic worlds but may have yet not been even put on a list of assumed objects.

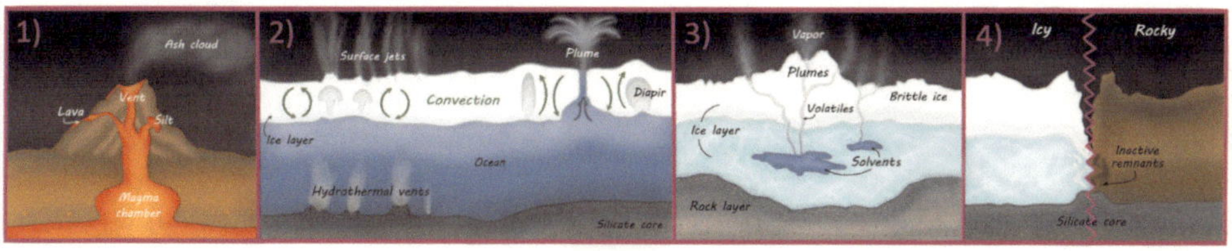

Figure 2.
Schematic overview of general types of volcanism (1–3) and how silicate and cryovolcanism are linked (2). Remnants of both silicate and cryovolcanism as signs of inactive volcanism in (4). Earth is a known example of silicate volcanism powered by accretion and radioactivity, as well as Io is also known example of silicate volcanism powered by tidal heating, may be both sketched in (1). Both known icy moons with cryovolcanism powered by tidal heating, Enceladus, and also triton may be found in (2) or in some parts may be in (3). Inactive remnants (4) as discussed may be found on many objects, for example, Vesta or the moon.

Considering this, we may, when looking out of our own solar system, get aware of how problematic identifications of volcanic worlds may get in these faraway systems. Also, some aspects may get stronger influence. Many systems with close orbits, favoring stronger tidal forces, especially around K- and M-stars, have been found and modeled (e.g., [114–118]). But many parameters of these systems being necessary for modeling are barely known and may need even stronger efforts in measuring and obtaining them. First attempts in reconsidering some constraints of these models have been done (as in e.g., [95]) and first assumptions based on reduced parameter sets for the evaluation of state and kind of volcanic worlds have been made. The approach aims at assessing the potential for volcanic worlds on easier than other observable parameters and has been verified in our own system, yielding all known and many assumed volcanic objects, plus hints for further bodies harboring volcanoes. Thus, it may be considered as a pre-scan before deeper and more intensive modeling. The first application in the system of TRAPPIST-1 gave rise to a higher volcanic potential on all planets, not only by forces of the central star but also by mutual tidal influences of the orbiting bodies [95].

4. Considerations for astrobiology

Regarding the phenomena of silicate and cryovolcanism, all of them may be powered by the energy sources discussed, but conditions for and evolution of these power sources are differing. Considering constraints for life as we know it, new aspects arise. Water in liquid form would be assumed as a requirement, in some alternative chemistry also ammonia or methane are discussed as possible solvents, liquid silicate/rock is less considered as being favorable for life. Also, a longer period of stability of these solvents is seen as favorable.

As accretion/radioactivity powered volcanism is high after formation and presumably gives rise to liquid silicates, it is a narrow gap of parameters depending on the size of the object and seeding of elements, which would allow a long and stable period of solvents as water. Bigger objects (starting already with radii just below 1000 km) might keep the heat over Gys too high, for example, water to rain down on the surface. Objects with sizes of several hundred kilometers and below may cool down very fast, allowing liquid water on the surface or in layers deeper in the crust for short periods of some 10 or 100 Mys [96]. Volcanism by tidal heating seems to be, if special conditions are met, more stable, as may be seen from all moons in our system with known active volcanism or tectonics, for example, Europa, Ganymede, or Enceladus. Even if becoming presumably unstable as Triton, it is after many Gys.

Considering the distribution of stable (e.g., considered from formation until now) volcanism powered by accretion/radioactivity or by tidal heating in our system, only Earth may be considered as accretion/radioactivity powered and many tens of objects powered by tidal heating confirmed or strongly assumed. If not for the power of the sun, habitable biotopes on Earth would be pretty much the same as the assumed ones on the moons discussed, that is, around vents deep in the liquid oceans below an ice crust covering (nearly) the whole surface. If we postulate such black smokers as life forging and maintaining harbors, in general, all over the universe, tidal heating may stably sustain such sources over many Gys, independent of a central stellar object even (and especially) on tiny objects. The requirements for tidal heating to power the cryovolcanism and rendering solvents liquid maybe not easily met, but considering the vast number of tiny objects (in contrast with bigger ones), the overall abundance of the self-powered systems may be seen as relatively high.

5. Conclusions

Silicate and cryovolcanism both occur in a broad spectrum considering the proofs, traces, and remnants in our own system. The constraints and challenges for detecting any volcanic activity beyond our system are huge. Some parameters maybe even far more difficult for measuring than others. Bigger objects with volcanism probably based mainly on accretion energy or radioactivity may still be easier for far distance observation, detection, and measurement. Still, an accompanying approach by modeling, for objects in our own system as well as beyond, based on measurable or other feasible attempts seems reasonable.

Considering the models and also the underlying energy sources and evolution, tidal heating as an energy source can be highly variable. It may have a broader spectrum in occurrence than heating by stored accretion energy or radioactivity. Tinier objects may get energy for significant heating from tidal heating and less from accretion and radioactivity. Objects may start in conditions for tidal heating, move out or in these conditions, and may be stabilized by accompanying partners. The real spectrum of possible sets of moons, asteroids, and planets will be probably even much broader. Considering the fact of much larger amounts of tiny objects, the implications for the probability of worlds with volcanic activity of any kind powered by tidal heating are huge.

Being aware of possible long stable periods for liquid solvents on such volcanic worlds powered by tidal heating and also considering known volcanic structures as deep ocean vents serving as harbors for genesis and maintenance of life, the relevance of tidal heating for cryovolcanism/low-temperature geological activity becomes even more prominent.

By a combination of observational systems and models, by their improvement and mutual influence, description and measurement of volcanic worlds, as well as possible biotopes for life beyond our own system, seems to be achievable.

Author details

Georg Hildenbrand[1*], Klaus Paschek[1,2], Myriam Schäfer[1] and Michael Hausmann[1]

1 Department of Physics and Astronomy, Heidelberg University, Kirchhoff-Institute for Physics, Heidelberg, Germany

2 Max Planck Institute for Astronomy, Heidelberg, Germany

*Address all correspondence to: hilden@kip.uni-heidelberg.de

References

[1] Williams DA, Howell RR. Active volcanis Effusive eruptions. In: Lopes RMC, Spencer JR, editors. Io Galileo New View Jupiter's Volcan. Moon, Berlin, Heidelberg: Springer; 2007. pp. 133-161. DOI: 10.1007/978-3-540-48841-5_7

[2] Volcanoes FP. A Planetary Perspective. Oxford, UK: Clarendon Press; 1993

[3] Williams DA, Byrne PK, Jozwiak L, Liu Y, Radebaugh J. 2 - Effusive silicate volcanis Observations and processes. In: TKP G, RMC L, Fagents SA, editors. Planetary Volcanism across the Solar System. Vol. 1 in Comparative Planetology. Amsterdam, The Netherlands: Elsevier; 2022. pp. 5-75. DOI: 10.1016/B978-0-12-813987-5.00002-X

[4] Zahnle K, Arndt N, Cockell C, Halliday A, Nisbet E, Selsis F, et al. Emergence of a habitable planet. Space Science Reviews. 2007;**129**:35-78. DOI: 10.1007/s11214-007-9225-z

[5] Head JW, Wilson L. Chapter 40 - volcanism on mercury. In: Sigurdsson H, editor. Encycl. Volcanoes. Second ed. Amsterda Academic Press; 2015. pp. 701-716. DOI: 10.1016/B978-0-12-385938-9.00040-7

[6] Braden SE, Stopar JD, Robinson MS, Lawrence SJ, van der Bogert CH, Hiesinger H. Evidence for basaltic volcanism on the moon within the past 100 million years. Nature Geoscience. 2014;**7**:787-791. DOI: 10.1038/ngeo2252

[7] Srivastava N, Kumar D, Gupta RP. Young viscous flows in the Lowell crater of Orientale basin, moon: Impact melts or volcanic eruptions? Planetary and Space Science. 2013;**87**:37-45. DOI: 10.1016/j.pss.2013.09.001

[8] Whitten J, Head JW, Staid M, Pieters CM, Mustard J, Clark R, et al. Lunar mare deposits associated with the Orientale impact basin: New insights into mineralogy, history, mode of emplacement, and relation to Orientale Basin evolution from moon mineralogy mapper (M^3) data from Chandrayaan-1. Journal of Geophysical Research, Planets. 2011;**116**:e00G09. DOI: 10.1029/2010JE003736

[9] Spudis PD. Chapter 39 - volcanism on the moon. In: Sigurdsson H, editor. Encycl. Volcanoes. Second ed. Amsterda Academic Press; 2015. pp. 689-700. DOI: 10.1016/B978-0-12-385938-9.00039-0

[10] Weber RC, Lin P-Y, Garnero EJ, Williams Q, Lognonné P. Seismic detection of the lunar Core. Science. 2011;**331**:309-312. DOI: 10.1126/science.1199375

[11] Dzurisin D. The tectonic and volcanic history of mercury as inferred from studies of scarps, ridges, troughs, and other lineaments. Journal of Geophysical Research - Solid Earth. 1978; **83**:4883-4906. DOI: 10.1029/JB083iB10p04883

[12] Hanson B. Mercury, up-close again. Science. 2008;**321**:58-58. DOI: 10.1126/science.321.5885.58

[13] Head JW, Chapman CR, Strom RG, Fassett CI, Denevi BW, Blewett DT, et al. Flood volcanism in the northern high latitudes of mercury revealed by MESSENGER. Science. 2011;**333**: 1853-1856. DOI: 10.1126/science.1211997

[14] Filiberto J, Trang D, Treiman AH, Gilmore MS. Present-day volcanism on Venus as evidenced from weathering rates of olivine. Science Advances. 2020; **6**(1):eaax7445. DOI: 10.1126/sciadv.aax7445. Available from: https://www.science.org/doi/abs/10.1126/sciadv.aax7445

[15] Shalygin EV, Markiewicz WJ, Basilevsky AT, Titov DV, Ignatiev NI, Head JW. Active volcanism on Venus in the Ganiki Chasma rift zone. Geophysical Research Letters. 2015;**42**:4762-4769. DOI: 10.1002/2015GL064088

[16] Armann M, Tackley PJ. Simulating the thermochemical magmatic and tectonic evolution of Venus's mantle and lithosphere: Two-dimensional models. Journal of Geophysical Research, Planets. 2012;**117**:E12003. DOI: 10.1029/2012JE004231

[17] Mikhail S, Heap MJ. Hot climate inhibits volcanism on Venus: Constraints from rock deformation experiments and argon isotope geochemistry. Physics of the Earth and Planetary Interiors. 2017; **268**:18-34. DOI: 10.1016/j.pepi.2017.05.007

[18] Hauber E, Brož P, Jagert F, Jodłowski P, Platz T. Very recent and wide-spread basaltic volcanism on Mars. Geophysical Research Letters. 2011;**38**: l10201. DOI: 10.1029/2011GL047310

[19] Horvath DG, Moitra P, Hamilton CW, Craddock RA, Andrews-Hanna JC. Evidence for geologically recent explosive volcanism in Elysium Planitia, Mars. Icarus. 2021;**365**:114499. DOI: 10.1016/j.icarus.2021.114499

[20] Fagents SA, Thordarson T. Rootless volcanic cones in Iceland and on Mars. In: Chapman MG, editor. Geol. Mars Evid. Earth-Based Analogs. Cambridge, UK: Cambridge University Press; 2007. pp. 151-177

[21] Keszthelyi LP, Jaeger WL, Dundas CM, Martínez-Alonso S, McEwen AS, Milazzo MP. Hydrovolcanic features on Mars: Preliminary observations from the first Mars year of HiRISE imaging. Icarus. 2010;**205**:211-229. DOI: 10.1016/j.icarus.2009.08.020

[22] Brož P, Hauber E. Hydrovolcanic tuff rings and cones as indicators for phreatomagmatic explosive eruptions on Mars. Journal of Geophysical Research, Planets. 2013;**118**:1656-1675. DOI: 10.1002/jgre.20120

[23] Chapman MG, Smellie JL. Mars interior layered deposits and terrestrial sub-ice volcanoes compared: Observations and interpretations of similar geomorphic characteristics. In: Chapman MG, editor. Geol. Mars Evid. Earth-Based Analog. Cambridge, UK: Cambridge University Press; 2007. pp. 178-207. DOI: 10.1017/CBO9780511536014.008

[24] Strom RG, Schaber GG, Dawson DD. The global resurfacing of Venus. Journal of Geophysical Research, Planets. 1994; **99**:10899-10926. DOI: 10.1029/94JE00388

[25] Byrne PK, Ghail RC, Şengör AMC, James PB, Klimczak C, Solomon SC. A globally fragmented and mobile lithosphere on Venus. Proceedings of the National Academy of Sciences. 2021;**118**: e2025919118. DOI: 10.1073/pnas.2025919118

[26] Mian Zu, Tozer Dc. No water, no plate tectonics: Convective heat transfer and the planetary surfaces of Venus and earth. Terra Nova. 1990;**2**:455-459. DOI: 10.1111/j.1365-3121.1990.tb00102.x

[27] Korenaga J. Plate tectonics and surface environment: Role of the oceanic upper mantle. Earth Science Reviews. 2020;**205**:103185. DOI: 10.1016/j.earscirev.2020.103185

[28] Schmandt B, Jacobsen SD, Becker TW, Liu Z, Dueker KG. Dehydration melting at the top of the lower mantle. Science. 2014;**344**: 1265-1268. DOI: 10.1126/science.1253358

[29] Zuber MT, McSween HY, Binzel RP, Elkins-Tanton LT, Konopliv AS, Pieters CM, et al. Origin, internal structure and evolution of 4 Vesta. Space Science Reviews. 2011;**163**:77-93. DOI: 10.1007/s11214-011-9806-8

[30] Moskovitz N, Gaidos E. Differentiation of planetesimals and the thermal consequences of melt migration. Meteoritics and Planetary Science. 2011; **46**:903-918. DOI: 10.1111/j.1945-5100.2011.01201.x

[31] Hansen CJ, Esposito L, Stewart AIF, Colwell J, Hendrix A, Pryor W, et al. Enceladus' water vapor plume. Science. 2006;**311**:1422-1425. DOI: 10.1126/science.1121254

[32] Soderblom LA, Kieffer SW, Becker TL, Brown RH, Cook AF, Hansen CJ, et al. Triton's geyser-like plumes: Discovery and basic characterization. Science. 1990;**250**: 410-415. DOI: 10.1126/science.250.4979.410

[33] Greenberg R, Geissler P, Hoppa G, Tufts BR. Tidal-tectonic processes and their implications for the character of Europa's icy crust. Reviews of Geophysics. 2002;**40**:1-33. DOI: 10.1029/2000RG000096

[34] Schenk PM, McKinnon WB, Gwynn D, Moore JM. Flooding of Ganymede's bright terrains by low-viscosity water-ice lavas. Nature. 2001; **410**:57-60. DOI: 10.1038/35065027

[35] De Waard D. Diapirism. In: Seyfert, editor. Struct. Geol. Tecton. Berlin, Heidelberg, Germany: Springer; 1987. pp. 202-203. DOI: 10.1007/3-540-31080-0_31

[36] Rajput S, Thakur NK. Chapter 4 - tectonics and gas hydrates. In: Rajput S, Thakur NK, editors. Geol. Controls Gas Hydrate Form. Unconv. Amsterdam, The Netherlands: Elsevier; 2016. pp. 107-130. DOI: 10.1016/B978-0-12-802020-3.00004-7

[37] Jet Propulsion Laboratory (JPL). California Institute of Technology (Caltech), National Araeronatics and Space Agengy (NASA). Solar System Bodies, JPL. n.d. https://ssd.jpl.nasa.gov/?bodies [Accessed April 11, 2022]

[38] Jet Propulsion Laboratory (JPL). California Institute of Technology (Caltech), National Araeronatics and Space Agengy (NASA). JPL Small-Body Database Search Engine. n.d. https://ssd.jpl.nasa.gov/sbdb_query.cgi [Accessed April 11, 2022]

[39] Ragozzine D, Brown ME. Orbits and masses of the satellites of the dwarf planet Haumea (2003 El61). Astronomy Journal. 2009;**137**:4766-4776. DOI: 10.1088/0004-6256/137/6/4766

[40] Brown ME, Schaller EL. The mass of dwarf planet Eris. Science. 2007; **316**:1585-1585. DOI: 10.1126/science.1139415

[41] Gillon M, Triaud AHMJ, Demory B-O, Jehin E, Agol E, Deck KM, et al. Seven temperate terrestrial planets around the nearby ultracool dwarf star TRAPPIST-1. Nature. 2017;**542**:456-460. DOI: 10.1038/nature21360

[42] Grimm SL, Demory B-O, Gillon M, Dorn C, Agol E, Burdanov A, et al. The nature of the TRAPPIST-1 exoplanets. Astronomy and Astrophysics. 2018;**613**: A68-A68. DOI: 10.1051/0004-6361/201732233

[43] Delrez L, Gillon M, Triaud AHMJ, Demory B-O, de Wit J, Ingalls JG, et al. Early 2017 observations of TRAPPIST-1 with Spitzer. Monthly Notices of the Royal Astronomical Society. 2018;**475**: 3577-3597. DOI: 10.1093/mnras/sty051

[44] Vilenius E, Kiss C, Mommert M, Müller T, Santos-Sanz P, Pal A, et al. "TNOs are cool": A survey of the trans-Neptunian region - VI. Herschel/PACS observations and thermal modeling of 19 classical Kuiper belt objects. Astronomy and Astrophysics. 2012;**541**:A94. DOI: 10.1051/0004-6361/201118743

[45] Dunham ET, Desch SJ, Probst L. Haumea's shape, composition, and internal structure. The Astrophysical Journal. 2019;**877**:41. DOI: 10.3847/1538-4357/ab13b3

[46] Ortiz JL, Santos-Sanz P, Sicardy B, Benedetti-Rossi G, Bérard D, Morales N, et al. The size, shape, density and ring of the dwarf planet Haumea from a stellar occultation. Nature. 2017;**550**:219-223. DOI: 10.1038/nature24051

[47] Braga-Ribas F, Sicardy B, Ortiz JL, Lellouch E, Tancredi G, Lecacheux J, et al. The size, shape, albedo, density, and atmospheric limit of Transneptunian object (50000) Quaoar from multi-chord stellar Occultations. The Astrophysical Journal. 2013;**773**:26. DOI: 10.1088/0004-637X/773/1/26

[48] Arimatsu K, Ohsawa R, Hashimoto GL, Urakawa S, Takahashi J, Tozuka M, et al. New constraint on the atmosphere of (50000) Quaoar from a stellar occultation. Astronomy Journal. 2019;**158**:236. DOI: 10.3847/1538-3881/ab5058

[49] Sicardy B, Ortiz JL, Assafin M, Jehin E, Maury A, Lellouch E, et al. Size, density, albedo and atmosphere limit of dwarf planet Eris from a stellar occultation. EPSC Abstracts. 2011; **2011**:137

[50] Saxena P, Renaud JP, Henning WG, Jutzi M, Hurford T. Relevance of tidal heating on large TNOs. Icarus. 2018;**302**: 245-260. DOI: 10.1016/j.icarus.2017.11.023

[51] Pál A, Kiss C, Müller TG, Santos-Sanz P, Vilenius E, Szalai N, et al. "TNOs are cool": A survey of the trans-Neptunian region - VII. Size and surface characteristics of (90377) Sedna and 2010 EK139. Astronomy and Astrophysics. 2012;**541**:L6. DOI: 10.1051/0004-6361/201218874

[52] Rommel FL, Braga-Ribas F, Desmars J, Camargo JIB, Ortiz JL, Sicardy B, et al. Stellar occultations enable milliarcsecond astrometry for trans-Neptunian objects and centaurs. Astronomy and Astrophysics. 2020;**644**: A40. DOI: 10.1051/0004-6361/202039054

[53] Grootel VV, Fernandes CS, Gillon M, Jehin E, Manfroid J, Scuflaire R, et al. Stellar parameters for Trappist-1. The Astrophysical Journal. 2018;**853**:30. DOI: 10.3847/1538-4357/aaa023

[54] Küppers M, O'Rourke L, Bockelée-Morvan D, Zakharov V, Lee S, von Allmen P, et al. Localized sources of water vapour on the dwarf planet (1) Ceres. Nature. 2014;**505**:525-527. DOI: 10.1038/nature12918

[55] McCord TB, Castillo-Rogez J, Rivkin A. Ceres: Its origin, evolution and structure and Dawn's potential

contribution. Space Science Reviews. 2011;**163**:63-76. DOI: 10.1007/s11214-010-9729-9

[56] Ruesch O, Platz T, Schenk P, McFadden LA, Castillo-Rogez JC, Quick LC, et al. Cryovolcanism on Ceres. Science. 2016;**353**:aaf4286. DOI: 10.1126/science.aaf4286

[57] Showman AP, Malhotra R. The Galilean satellites. Science. 1999;**286**: 77-84. DOI: 10.1126/science.286.5437.77

[58] Anderson JD, Schubert G, Jacobson RA, Lau EL, Moore WB, Sjogren WL. Europa's differentiated internal structure: Inferences from four Galileo encounters. Science. 1998;**281**: 2019-2022. DOI: 10.1126/science.281.5385.2019

[59] Fagents SA. Considerations for effusive cryovolcanism on Europa: The post-Galileo perspective. Journal of Geophysical Research, Planets. 2003;**108** (e12):5139. DOI: 10.1029/2003JE002128

[60] Quick LC, Glaze LS, Baloga SM. Cryovolcanic emplacement of domes on Europa. Icarus. 2017;**284**:477-488. DOI: 10.1016/j.icarus.2016.06.029

[61] Vance S, Bouffard M, Choukroun M, Sotin C. Ganymede's internal structure including thermodynamics of magnesium sulfate oceans in contact with ice. Planetary and Space Science. 2014;**96**:62-70. DOI: 10.1016/j.pss.2014.03.011

[62] Lee Allison M, Clifford SM. Ice-covered water volcanism on Ganymede. Journal of Geophysical Research - Solid Earth. 1987;**92**:7865-7876. DOI: 10.1029/JB092iB08p07865

[63] Tajeddine R, Rambaux N, Lainey V, Charnoz S, Richard A, Rivoldini A, et al. Constraints on Mimas' interior from Cassini ISS libration measurements. Science. 2014;**346**:322-324. DOI: 10.1126/science.1255299

[64] Rhoden AR, Henning W, Hurford TA, Patthoff DA, Tajeddine R. The implications of tides on the Mimas Ocean hypothesis. Journal of Geophysical Research, Planets. 2017;**122**: 400-410. DOI: 10.1002/2016JE005097

[65] Rhoden AR, Walker ME. The case for an ocean-bearing Mimas from tidal heating analysis. Icarus. 2022;**376**: 114872. DOI: 10.1016/j.icarus.2021.114872

[66] Beuthe M, Rivoldini A, Trinh A. Enceladus's and Dione's floating ice shells supported by minimum stress isostasy. Geophysical Research Letters. 2016;**43**:10,088-10,096. DOI: 10.1002/2016GL070650

[67] Chen EMA, Nimmo F. Implications from Ithaca Chasma for the thermal and orbital history of Tethys. Geophysical Research Letters. 2008;**35**:l19203. DOI: 10.1029/2008GL035402

[68] Hussmann H, Rodríguez A, Callegari N, Shoji D. Early resonances of Tethys and Dione: Implications for Ithaca Chasma. Icarus. 2019;**319**: 407-416. DOI: 10.1016/j.icarus.2018.09.025

[69] Gyalay S, Dodds KH, Nimmo F. Estimates of Tethys' present-day heat flux and moment of inertia from its long-wavelength topography. AGU Fall Meeting Abstracts. 2018;**2018**:P54B-P507B

[70] Hussmann H, Sohl F, Spohn T. Subsurface oceans and deep interiors of medium-sized outer planet satellites and large trans-neptunian objects. Icarus. 2006;**185**:258-273. DOI: 10.1016/j.icarus.2006.06.005

[71] Owen T. Huygens rediscovers titan. Nature. 2005;**438**:756-757. DOI: 10.1038/438756a

[72] Grasset O, Sotin C, Deschamps F. On the internal structure and dynamics of titan. Planetary and Space Science. 2000; **48**:617-636. DOI: 10.1016/S0032-0633(00)00039-8

[73] Sotin C, Jaumann R, Buratti BJ, Brown RH, Clark RN, Soderblom LA, et al. Release of volatiles from a possible cryovolcano from near-infrared imaging of Titan. Nature. 2005;**435**:786-789. DOI: 10.1038/nature03596

[74] Turtle EP, Perry JE, McEwen AS, DelGenio AD, Barbara J, West RA, et al. Cassini imaging of Titan's high-latitude lakes, clouds, and south-polar surface changes. Geophysical Research Letters. 2009;**36**:l02204. DOI: 10.1029/2008GL036186

[75] Radebaugh J, Lorenz RD, Kirk RL, Lunine JI, Stofan ER, Lopes RMC, et al. Mountains on titan observed by Cassini Radar. Icarus. 2007;**192**:77-91. DOI: 10.1016/j.icarus.2007.06.020

[76] Lopes RMC, Kirk RL, Mitchell KL, LeGall A, Barnes JW, Hayes A, et al. Cryovolcanism on titan: New results from Cassini RADAR and VIMS. Journal of Geophysical Research, Planets. 2013; **118**:416-435. DOI: 10.1002/jgre.20062

[77] Wood CA, Radebaugh J. Morphologic evidence for volcanic craters near Titan's north polar region. Journal of Geophysical Research, Planets. 2020;**125**:e2019JE006036. DOI: 10.1029/2019JE006036

[78] Fortes AD, Grindrod PM, Trickett SK, Vočadlo L. Ammonium sulfate on titan: Possible origin and role in cryovolcanism. Icarus. 2007;**188**: 139-153. DOI: 10.1016/j.icarus.2006.11.002

[79] Mitri G, Bland MT, Showman AP, Radebaugh J, Stiles B, Lopes RMC, et al. Mountains on titan: Modeling and observations. Journal of Geophysical Research, Planets. 2010;**115**:e10002. DOI: 10.1029/2010JE003592

[80] Ring Around a Moon? n.d. https://www.science.org/doi/10.1126/science.307.5708.349c [accessed: March 27, 2022]

[81] Tittemore WC, Wisdom J. Tidal evolution of the Uranian satellites: III. Evolution through the Miranda-Umbriel 3:1, Miranda-Ariel 5:3, and Ariel-Umbriel 2:1 mean-motion commensurabilities. Icarus. 1990;**85**:394-443. DOI: 10.1016/0019-1035(90)90125-S

[82] Schenk PM. Fluid volcanism on Miranda and Ariel: Flow morphology and composition. Journal of Geophysical Research - Solid Earth. 1991;**96**: 1887-1906. DOI: 10.1029/90JB01604

[83] Pappalardo RT, Reynolds SJ, Greeley R. Extensional tilt blocks on Miranda: Evidence for an upwelling origin of Arden Corona. Journal of Geophysical Research, Planets. 1997;**102**: 13369-13379. DOI: 10.1029/97JE00802

[84] Hammond NP, Barr AC. Global resurfacing of Uranus's moon Miranda by convection. Geology. 2014;**42**: 931-934. DOI: 10.1130/G36124.1

[85] Martin-Herrero A, Romeo I, Ruiz J. Heat flow in triton: Implications for heat sources powering recent geologic activity. Planetary and Space Science. 2018;**160**:19-25. DOI: 10.1016/j.pss.2018.03.010

[86] Kargel JS. Cryovolcanism on the icy satellites. Earth, Moon, and Planets.

1994;**67**:101-113. DOI: 10.1007/BF00613296

[87] Smith BA, Soderblom LA, Banfield D, Barnet C, Basilevsky AT, Beebe RF, et al. Voyager 2 at Neptune: Imaging science results. Science. 1989; **246**:1422-1449. DOI: 10.1126/science.246.4936.1422

[88] McKinnon WB, Kirk RL. Triton. In: Johnson TV, Spohn T, Breuer D, editors. Encycl. Sol. Syst. 3rd ed., Amsterdam; Boston: Elsevier; 2014, p. 861–882.

[89] Strom RG, Croft SK, Boyce JM. The impact cratering record on triton. Science. 1990;**250**:437-439. DOI: 10.1126/science.250.4979.437

[90] Witze A. Icy volcanoes may dot Pluto's surface. Nature. 2015. DOI: 10.1038/nature.2015.18756

[91] Desch SJ, Cook JC, Doggett TC, Porter SB. Thermal evolution of Kuiper belt objects, with implications for cryovolcanism. Icarus. 2009;**202**: 694-714. DOI: 10.1016/j.icarus.2009.03.009

[92] Barucci MA, Merlin F, Guilbert A, Bergh C de, Alvarez-Candal A, Hainaut O, et al. Surface composition and temperature of the TNO Orcus. Astronomy and Astrophysics 2008;479: L13–L16. DOI: 10.1051/0004-6361:20079079

[93] Dumas C, Carry B, Hestroffer D, Merlin F. High-contrast observations of (136108) Haumea - a crystalline water-ice multiple system. Astronomy and Astrophysics. 2011;**528**:A105. DOI: 10.1051/0004-6361/201015011

[94] Jewitt DC, Luu J. Crystalline water ice on the Kuiper belt object (50000) Quaoar. Nature. 2004;**432**:731-733. DOI: 10.1038/nature03111

[95] Paschek K, Roßmann A, Hausmann M, Hildenbrand G. Analysis of tidal accelerations in the solar system and in extrasolar planetary systems. Applied Sciences. 2021;**11**:8624. DOI: 10.3390/app11188624

[96] Guilbert-Lepoutre A, Lasue J, Federico C, Coradini A, Orosei R, Rosenberg ED. New 3D thermal evolution model for icy bodies application to trans-Neptunian objects. Astronomy and Astrophysics. 2011;**529**: A71. DOI: 10.1051/0004-6361/201014194

[97] Efroimsky M, Makarov VV. Tidal dissipation in a homogeneous spherical body. International Methods Astrophysics Journal. 2014;**795**:6. DOI: 10.1088/0004-637X/795/1/6

[98] Zschau J. Tidal friction in the solid earth: Loading tides versus body tides. In: Brosche P, Sündermann J, editors. Tidal Frict. Earth's Rotat. Berlin, Heidelberg, Germany: Springer Berlin Heidelberg; 1978. pp. 62-94. DOI: 10.1007/978-3-642-67097-8_7

[99] Platzman GW. Planetary energy balance for tidal dissipation. Reviews of Geophysics. 1984;**22**:73-84. DOI: 10.1029/RG022i001p00073

[100] Segatz M, Spohn T, Ross MN, Schubert G. Tidal dissipation, surface heat flow, and figure of viscoelastic models of Io. Icarus. 1988;**75**:187-206. DOI: 10.1016/0019-1035(88)90001-2

[101] Peale SJ. Generalized Cassini's Laws. Astronomy Journal. 1969;**74**:483. DOI: 10.1086/110825

[102] Peale SJ, Cassen P. Contribution of tidal dissipation to lunar thermal history. Icarus. 1978;**36**:245-269. DOI: 10.1016/0019-1035(78)90109-4

[103] Kaula WM. Theory of Satellite Geodesy. Applications of Satellites to Geodesy. Waltham, MA, USA: Blaisdell Publishing Company; 1966

[104] Bills BG, Neumann GA, Smith DE, Zuber MT. Improved estimate of tidal dissipation within Mars from MOLA observations of the shadow of Phobos. Journal of Geophysical Research, Planets. 2005;**110**:e07004. DOI: 10.1029/2004JE002376

[105] Taylor PA, Margot J-L. Tidal evolution of close binary asteroid systems. Celestial Mechanics and Dynamical Astronomy. 2010;**108**: 315-338. DOI: 10.1007/s10569-010-9308-0

[106] Goldreich P, Soter S. Q in the solar system. Icarus. 1966;**5**:375-389. DOI: 10.1016/0019-1035(66)90051-0

[107] Gladman B, Quinn DD, Nicholson P, Rand R. Synchronous locking of tidally evolving satellites. Icarus. 1996;**122**:166-192. DOI: 10.1006/icar.1996.0117

[108] Peale SJ. Rotation histories of the natural satellites. In: Burns JA, editor. Proc. IAU Colloq 28 Ithaca NY USA. Tucson, AZ, USA: University of Arizona Press; 1977. p. 87

[109] Lainey V, Arlot J-E, Karatekin Ö, Van Hoolst T. Strong tidal dissipation in Io and Jupiter from astrometric observations. Nature. 2009;**459**:957-959. DOI: 10.1038/nature08108

[110] Yoder CF. How tidal heating in Io drives the galilean orbital resonance locks. Nature. 1979;**279**:767-770. DOI: 10.1038/279767a0

[111] Nobili AM. Secular effects of tidal friction on the planet-satellite systems of the solar system. Moon and the Planets. 1978;**18**:203-216. DOI: 10.1007/BF00896743

[112] Ćuk M, Gladman BJ. Constraints on the orbital evolution of triton. The Astrophysical Journal. 2005;**626**: L113-L116. DOI: 10.1086/431743

[113] Chyba CF, Jankowski DG, Nicholson PD. Tidal evolution in the Neptune-Triton system. Astronomy and Astrophysics. 1989;**219**:L23-L26

[114] Kite ES, Manga M, Gaidos E. Geodynamics and rate of volcanism on massive earth-like planets. The Astrophysical Journal. 2009;**700**: 1732-1749. DOI: 10.1088/0004-637X/700/2/1732

[115] Barr AC, Dobos V, Kiss LL. Interior structures and tidal heating in the TRAPPIST-1 planets. Astronomy and Astrophysics. 2018;**613**:A37. DOI: 10.1051/0004-6361/201731992

[116] Dobos V, Barr AC, Kiss LL. Tidal heating and the habitability of the TRAPPIST-1 exoplanets. Astronomy and Astrophysics. 2019;**624**:A2. DOI: 10.1051/0004-6361/201834254

[117] Bolmont E, Breton SN, Tobie G, Dumoulin C, Mathis S, Grasset O. Solid tidal friction in multi-layer planets: Application to earth, Venus, a super earth and the TRAPPIST-1 planets - potential approximation of a multi-layer planet as a homogeneous body. Astronomy and Astrophysics. 2020;**644**:A165. DOI: 10.1051/0004-6361/202038204

[118] Bolmont E, Selsis F, Raymond SN, Leconte J, Hersant F, Maurin A-S, et al. Tidal dissipation and eccentricity pumping: Implications for the depth of the secondary eclipse of 55 Cancri e. Astronomy and Astrophysics. 2013;**556**: A17. DOI: 10.1051/0004-6361/201220837

Chapter 6

Is the Ocean of Enceladus in a Primitive Evolutionary Stage?

Katherine Villavicencio Valero, Emilio Ramírez Juidías and Aina Àvila Bosch

Abstract

Enceladus has a subsurface ocean in the South Pole that has been inferred due to the presence of water vapor and other molecules like molecular hydrogen and ammonia detected by the Cassini mission from the ejection of material through the plumes in that region. The chemical composition of this ocean could give some information about the evolutionary stage of the icy moon if its components are found to be similar with the aqueous chemistry of the primitive oceans on Earth during glacial periods. Here we present a comparative geochemical analysis between the ocean of Enceladus and the aqueous composition of the oceans on Earth during the Snowball Event, in order to figure out if there are similar species, how the interaction of the metabolic processes between them works and if, in the future, those molecules could evolve making possible the emergence of life.

Keywords: ocean, snowball event, aqueous chemistry, species, life

1. Introduction

Enceladus, one of the moons of Saturn, presents a global ocean beneath the ice shell [1]. The existence of that ocean was suggested because of the water vapor detected by the Cassini mission, through the ejection of material from the plumes located in the south pole [2, 3]. The expulsion of material from the water plumes could be related to hydrothermal activity [4], where ice particles are heated due to the tidal deformation [5] and expelled to the surface. Evidence that those particles are associated to hydrothermal activity are the silicate salts residues found at the E-ring [6], and the small size of the nanoparticles of that ring. Both characteristics indicate that the possible liquid water within the ocean layer was previously in contact with a hot silicate environment [7].

The Ion and Neutral Mass Spectrometer (INMS) instrument on board of the Cassini mission also detected ammonia and some traces of organic molecules like benzene [8]. Ammonia is one more clue of the presence of liquid water. Residuals from ammonia are nitrogen-bearing and oxygen-bearing molecules that, in combination, could convert into amino acids like it happens on Earth [9]. Other detected species were H_2O and CO_2 [10]. The metabolic interaction between these latter two species, through methanogenesis, can form methane. It was also detected molecular hydrogenH_2 by the Cassini spacecraft [11]. There were also found some species with compounds of carbon, nitrogen, oxygen and sulfur [8, 12, 13]. The interaction between molecular hydrogen and some carbonates within the ocean produce a chemical instability that constitutes an energy source that may support life [11].

The Cosmic Dust Analyzer (CDA) aboard of the Cassini mission detected water ice, organic molecule, and siliceous material [14]. There were also detected concentrations of Na, and some sodium salts like NaCl, $NaHCO_3$ and Na_2CO_3 indicating the presence of liquid water [15, 16]. To maintain this liquid water into the global ocean, the tidal dissipation could be considered as an energy source that come from inside Enceladus [17]. Tidal heating also acts in the solid core provoking high temperatures into the hydrothermal activity [18]. The hydrothermal activity creates convection columns that produce a dynamic movement in the ocean transporting the heat into the ice shell from the core [19]. The dissipation of the heat linked to the gas ratios present in the water plume could determine the state of the hydrothermal activity [7].

According to Woods [20], a hydrothermal source of gas could explain the distribution of hydrogen in the water plume. In this sense, it must be emphasized that the abundance of hydrogen detected is similar to some traces of volatile compounds like carbon dioxide, methane, and ammonia [8]. Laboratory simulations [7] suggested that, in Enceladus, the molecular hydrogen is a product of internal reactions. Evidence of the internal production of molecular hydrogen is the high ratio of H_2/H_2O which cannot come from a gas trapped in the ocean, because of its high concentration, and it cannot also come from the remnants of a formation environment, due to the low ratio ofHe/H_2 [21].

The geochemical system of the ocean of Enceladus could be composed mainly byNa_2O -HCl-CO_2 -H_2O. The concentration of CO_2 in the plumes is assumed to be the same that may be found in the dissolved molecules of the subsurface ocean. Its presence suggests a basic pH for the ocean of Enceladus. The estimation of this pH is based on the study of the thermodynamic equilibrium, considering the temperature close to 0°C, the pressure at 1 bar, the carbon dioxide activity, the chloride concentration, and the dissolved inorganic carbon HCO_3^-/CO_3^{2-}. According to Glein et al. [22], the pH is 12.15 ± 1.15.

Carbonates and bicarbonates ions $CO_3^{2-}/NaCO_3^-$ are also present in the ocean [23] and could come from soluble carbonate minerals, formed through the reaction of a trapped CO_2 and silica minerals during water-rock differentiation. If the rocks of the ocean react withCO_2, it is feasible the carbonation process for high water-rock ratios [24]. The metal concentration in the seawater is formed by phyllosilicates and hydroxide minerals, that need an acid in order to hydrolyze and being incorporated into carbonate minerals [23].

CO_2/H_2O ratio in the plume of Enceladus is similar to the ratio found in the seawater on Earth [23], where theCO_2 activity is controlled by alteration of minerals, assuming that water-rock interactions are the main driving force of the pH as well as the composition of the ocean. On Enceladus, the serpentinization is assumed to be the result of minerals alteration [25], through a hydrolysis of primary minerals containing iron and magnesium which product is the hydrogen. This process is usually associated with ultramafic rocks (<45% of SiO_2 and high Mg - Fe content), which reaction that takes place is the oxidation by water of Fe(II) and Fe(III) in minerals such as olivine and pyroxene. The product of this reaction is the molecular hydrogen.

The presence of hydrogen can form linear chains of hydrocarbons like methane CH_4 from the chemical reaction between CO_2 and H_2. This methane could be present in low concentrations into the ocean of Enceladus [26]. Evidence of this, it is the formation of clathrates that are able to trap certain molecules, which then, would rise to the surface and eventually dissociate and enrich the plume with methane. Methanol was also detected by the Cassini mission; it is possible that this compound has a biological origin [27]. It was found that the CH_3OH/H_2O ratio has certain correlation with biotic activity around the hydrothermal vents. The concentration of methanol detected in the atmosphere is high, which gives a clue about that this specie is formed beneath the ice shell before being expelled into the atmosphere. These organic compounds detected could be considered as a building block of life or even by-products of life [28].

On Earth, the first signs of life came from the Archean oceans where the oxidative reactions were a product of the interaction between molybdenum and rhenium [29]. There were only traces of oxygen before the Great Oxygen Event but then, after it, the photosynthetic activity led to an increment of this element [30]. The evolution of oxygen in the atmosphere and oceans went through five stages [31]. During the Cryogenian age, the atmosphere and the shallow oceans had an increase of oxygen. The oxygen concentration was stagnant in that era, and subsequently it had an increment that continued after the next million years and might have culminated around the Carboniferous age. During glacial periods, the concentration of CO_2 in the atmosphere dropped and, before the emergence of photosynthetic life, the carbon dioxide was more abundant in the atmosphere than nowadays [32].

Abundance of CO_2 in the atmosphere during that time would be a consequence of a carbon-silicate cycle during millions of years that after changed the Snowball events conditions [33]. The concentration of oxygen was low in the oceans during the Snowball periods. The water had a high level of acidity due to the high concentration of CO_2 in the atmosphere [32]. The ice-covered conditions on Earth were altered because of the melting of the ice crust that took place due to the increase of the temperature by volcanoes activity, which reduced the presence of CO_2 in the atmosphere and provoked the emergence of liquid water [34]. The high volcanic activity triggered the extensive presence of hydrothermal vents during the Cryogenian age [35–37].

Nowadays, hydrothermal systems can be classified as black smokers and lost city systems. The first one, are characterized by the black smoke that rises from the chimney-like rocky formations, where seawater is in contact with the magma chambers and emerges with an acid pH 2–3, a high content of dissolved metals such as Fe (II) and Mn (II), a variety of gases originated from volcanic activity like CO_2, H_2S, H_2, CH_4 and also with high temperatures up to 405°C. In contrast, in the lost city systems, the water that circulates trough the vents is not in contact with the magma, instead, it is heated by convection from the mantle and by exothermic chemical reactions between the fluid and the surrounding rocks reaching temperatures of 200°C [35]. The rock that interacts with the fluid is dominated by low-silica iron and magnesium rich minerals, provoking the methanogenesis by serpentinization of the hydrogen and the reduction of carbon dioxide in the ocean. In this case, the pH of the fluid is basic 9–11, it has dissolved gases likeH_2, CH_4, low-mass of hydrocarbons, and a low dissolved CO_2.

Similarities could be found along with the ancient oceans on Earth during the Snowball Events and the current conditions of the ocean on Enceladus. Here we present a comparative geochemistry analysis of both oceans. We also describe a chemical metabolic process based on numerical simulations that could take place within the global ocean of Enceladus, in order to infer if the current conditions of that ocean could evolve to create the building chains of life. During glaciations ages, the ice-covered Earth allowed for maintaining the liquid water beneath the ice crust, and subsequently that liquid water emerged to the surface by the hot spots or hydrothermal vents once the high concentration of CO_2 started changing the conditions of the atmosphere [38]. On Enceladus, there are hints that indicate the presence of liquid water, such as the hydrated sodium salts detected by the Cassini mission. The molecular hydrogen found also gives clues about a hydrothermal activity beneath the ice shell. We aim to infer a possible evolutionary stage of the ocean of Enceladus that could make possible the emergence of life.

2. Study area

Models of the internal structure of Enceladus reveals an ocean on average 26–31 km in depth below an ice layer between 21 and 26 km of thickness [39]. Salinity geochemistry simulations of the ocean of Enceladus show values nearly similar to Earth, around 20 g/kg [40, 41]. The quantity of water vapor ejected from the plumes is around 150–300 kg/s [42]. This ejection of particles supply the composition of the ring E of Saturn, with <10% of the material catching into it, also suggesting a liquid origin [43, 44]. **Figure 1** shows the distribution of the plumes along the south pole of Enceladus.

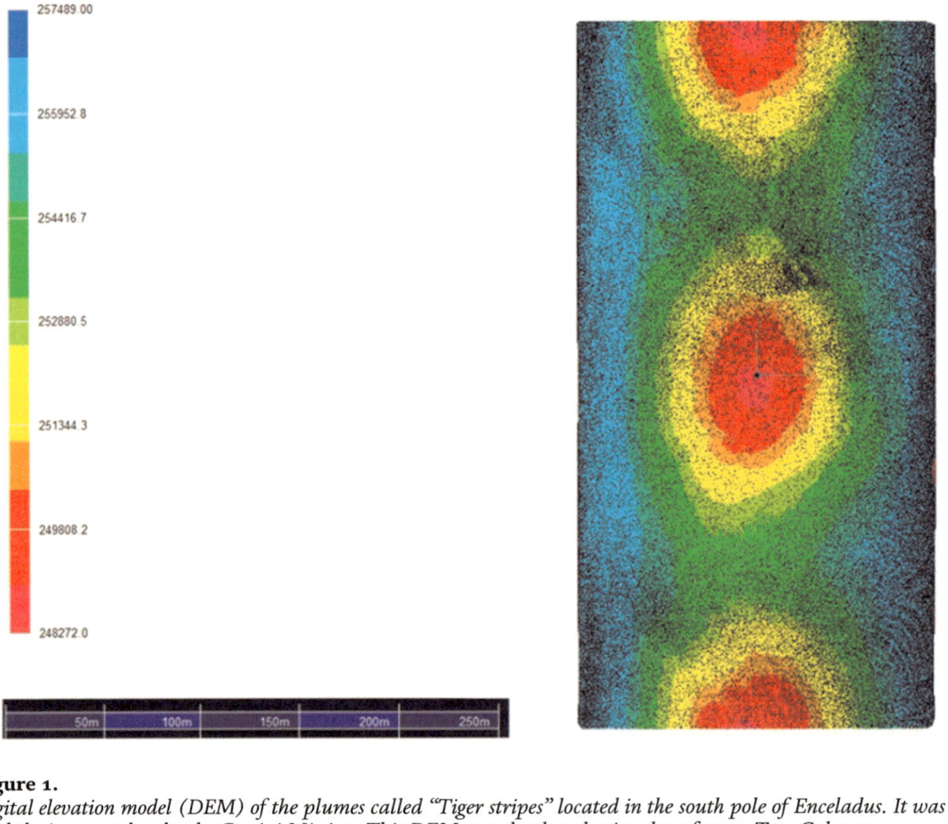

Figure 1.
Digital elevation model (DEM) of the plumes called "Tiger stripes" located in the south pole of Enceladus. It was used the images taken by the Cassini Mission. This DEM was developed using the software TopoCal 2022 v.9.0.811.

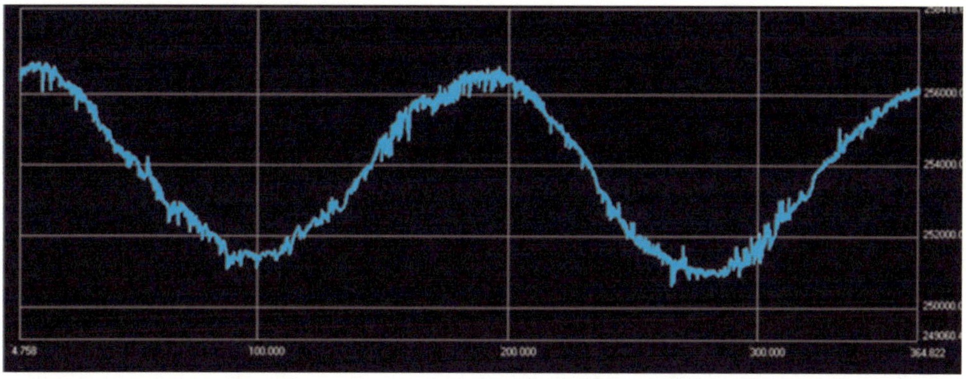

Figure 2.
Elevation profile of the DEM from top to bottom.

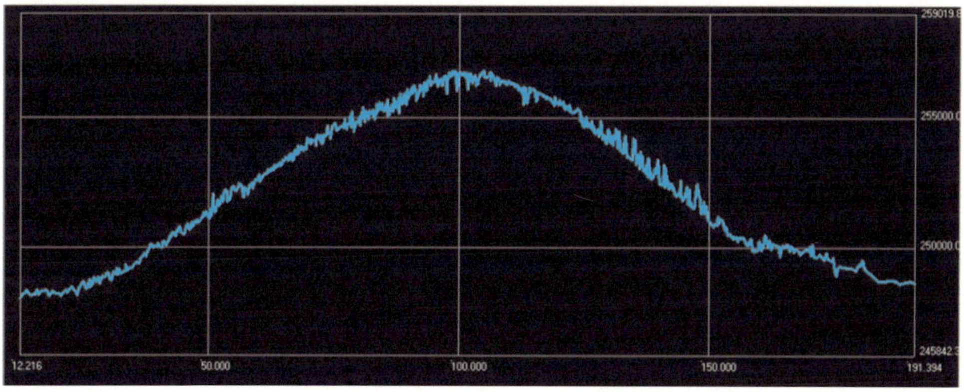

Figure 3.
Elevation profile of the DEM to the center from left to right.

Beneath the south pole the composition of the particles is mainly salt rich, implying that those salts are larger than salt-poor grains and they are expelled with lower escape velocity. The escape speed of particles from the plumes in Enceladus is on average 1.85–2.25 km/s, according to the measures from the dusty plume by the flyby of the Cassini spacecraft [45]. **Figures 2** and **3** show the longitudinal (y axis) and transversal (x axis) height profiles of the plumes of Enceladus from **Figure 1**. The longitudinal axis of **Figure 2** presents a radius in the central plume of 190 m, besides, the transversal axis of **Figure 3** shows a radius of 90 m. The distribution of the fissures along the plumes seems to be aligned in the y axis.

3. Materials and methods

In this research, we used the data of the molecular species detected by the INMS instrument on board of the Cassini Mission. The spectral signatures were encoded according to Ramírez-Juidías et al. [46]. Then, there were selected the common spectral lines present in the ocean of Enceladus and in the seawater of the oceans on Earth. Based on the spectral lines, it was applied data mining in order to extract the

Species	Enceladus concentration (g/kg)	Earth concentration (g/kg)
$B(OH)_{4^-}$		0,008
$B(OH)_3$		0,019
Br		0,067
Ca^{2+}		0,412
Cl^-	7076	19,353
$CO_3{}^{2-}$	2867	0,016
F^-		0,013
$HCO_3{}^-$	0,031	0,107
K^+		0,399
Mg^{2+}		1284
Na^+	7343	10,784
NaCl	0,024	
$NaCO_3{}^-$	1788	
$NaHCO_3$	0,015	
NaOH	0,008	
OH^-	0,038	
$SO_4{}^{2-}$	0,1–0,01	2713
Sr^{2+}		0,008

Table 1.
Concentration of species in the ocean of Enceladus and in the seawater of the oceans on earth.

concentration of species detected in the material ejected from the plumes. **Table 1** shows some of the species present in the ocean of Enceladus and the seawater of the oceans on Earth [47] with their concentration in g/kg. Each specie was extrapolated to the geochemical processes associated to the activity of CO_2 and H_2O within the ocean [22].

According to the method patented by Ramírez-Juidías et al. [46], the data mining process was carried out through the application of modified genetic algorithms, iteratively analyzing a large amount of data through a process similar to genetic mutation, in order to extract the variables that are then used to obtain the concentrations (g/kg) of species in the ocean of Enceladus, using the wavelengths between 0.35 and 1 μm from the spectral data taken by the VIMS instrument.

The encoding model developed to obtain these concentrations consists in building a vector of size equals to the number of iterations to execute. The kth-order of the vector represents the work that is done in the kth-position. In this case, a population of alternative solutions is settled for a certain number of chromosomes, that represent the natural sequence in which the variables (spectral signatures) are programmed.

The process of planning and programming required for the extraction of the concentrations of species is usually conducted by applying a three-level model called respectively Strategic Approach, Tactical Approach and Operational Approach. This model can be replicated using machine learning.

4. Results and discussion

Sodium ion and Chlorine are the most abundant species in the ocean of Enceladus. **Figure 4** shows the concentration of both species in mol per kg of H_2O. The quantity of mol is calculated in function of the CO_2 activity. Na^+ is present in concentrations around 0.800 mol/kg while the concentrations of Cl^- are constant, around 0.400 mol/kg. **Figure 5** also shows that there are present some carbonates CO_3^{2-}, bicarbonate HCO_3^- and sulfate SO_4^{2-}. CO_3^{2-} has a concentration on average 0.075–0.030 mol/kg which tends to decrease with the activity of CO_2. HCO_3^- presents an increment from 0.020 to 0.070 mol/kg and SO_4^{2-} has a concentration between 0.01 and 0.1 mol/kg.

The concentrations of salinity and chlorinity are relatively constant in the current terrestrial oceans. The average concentration from the seawater with a pH of 8.1 and temperature of 25°C are detailed in **Table 1**. Geological information extracted from sedimentary layers reveals that deep oceans were in a reduced state till the end of the Paleoproterozoic era. Iron and calcium sulfate probably played as reduced agents with the oxygen converting FeO into Fe_2O_3, and precipitating $CaSO_4$. During the

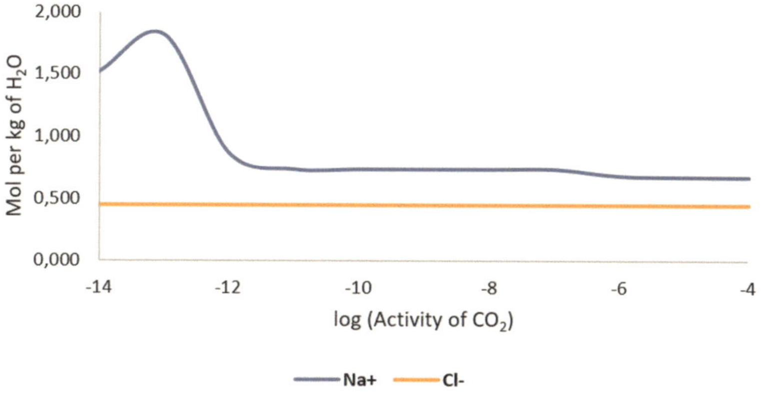

Figure 4.
Sodium ion and chlorine present in the ocean of Enceladus. The concentration of species are calculated in function of the CO_2 *activity.*

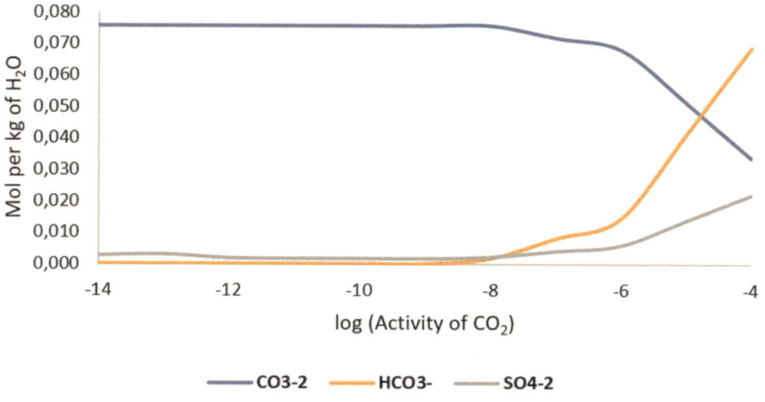

Figure 5.
Some carbonates, bicarbonates, and sulfate present in the ocean of Enceladus. The concentration of species are calculated in function of the CO_2 *activity.*

Species	Enceladus concentration (g/kg)	Earth concentration (g/kg)
Cl^-	7076	19,353
CO_3^{2-}	2867	0,016
HCO_3^-	0,031	0,107
Na^+	7343	10,784
SO_4^{2-}	0,1 - 0,01	2713

Table 2.
Concentration of species present in the seawater of the oceans on earth and in the ocean of Enceladus.

Cryogenian era the concentration of sulphate could rise to levels similar to the recent ones, around 23 mol/kg of H_2O [47].

Table 2 shows few key species present in the ocean of Enceladus and in the seawater of the oceans on Earth. Sodium ion and Chlorine are the most abundant species in both oceans. The oceans on Earth are saltier with a pH of 8.1 on average, while the ocean of Enceladus is more basic, around pH 12.2. The ocean of Enceladus has more dissolved inorganic carbon than the ocean on Earth. On Enceladus, the predominant carbonate is CO_3^{2-} while on Earth is the bicarbonateHCO_3^-. The abundance of CO_3^{2-}in the ocean of Enceladus could be due to the serpentinization of the molecular hydrogen. The concentration of sulfur in the ocean of Enceladus is variable compared to the one present on Earth.

Two scenarios can be considered to calculate the amount of sulphate that could be oxidized on the ocean of Enceladus. The lower concentration of SO_4^{2-}, 0.01 g/kg, displayed in **Table 2**, takes place only in aqueous reductants environments where HS^- reacts with the oxidants, while in the larger concentration 0.1 g/kg, some minerals are considered as a source for reductants. The concentration of sulphate in the ocean of Enceladus is below to the current amount of sulphate found on the oceans on Earth but, this concentration could have been smaller during the Snowball events, being close to the current quantities on the ocean of Enceladus.

The predominant concentration of inorganic carbonate species found in the ocean of Enceladus, set the ocean as not compatible with life except for the methane detected that can be a product of the methanogenesis of the carbon dioxide and the hydrogen. Would it be possible that the species detected in the ocean of Enceladus evolve to create the chains of life? how were the chemical conditions of the primitive terrestrial oceans before rising life? In order to figure out which similarities could be found between the terrestrial oceans and the ocean of Enceladus, it is necessary to understand the evolution of the ancient aqueous geochemistry of the oceans in the primitive Earth.

During the first stage of formation of Earth, it was bombarded by hydrous asteroids mainly type Cl chondrites bringing water, organic molecules, and chondritic minerals. Tectonic activity facilitated to diversity the mineralogy along the crust, increasing the mafic content of the top layers through the eruption of hot basaltic lavas. Chondritic material has been also detected in the plumes of Enceladus [14, 21], that is why, it could be possible to infer that this material can be settled in the seafloor of its ocean [48].

Organisms cannot devise chemical processes by themselves, they must copy natural reactions, adapt them, and optimize them through time. Phosphorylation is the addition of a phosphate group into a protein, being the main mechanism of biochemistry. This mechanism participates in some proteins regulation like ATP formation, fatty acids metabolization, and citric acid cycle. Prebiotic phosphorylation of biological molecules is a reaction that represent a challenge for the study of the origin of life. It has been proven that using diamidophosphate (DAP) instead of phosphates, thermodynamic barriers decreased for this reaction in water, and different organic building blocks were able to be assembled [49]. Based on that analysis, it was demonstrated that is possible to generate DAP and other amino - phosphor compounds when P-bearing molecules are mixed with aqueous ammonia solutions. The sources of phosphor could come from iron P-bearing minerals, condensed phosphates which contain salts and metals, or reduced phosphorus compounds. Those reactions probably took place in the aqueous conditions of the early Earth. If the concentration of ammonia in the hydrothermal vents of Enceladus would be similar to the prebiotic oceans on Earth, that phosphate reaction could happen in the ocean of Enceladus.

Although the currents anaerobic sulfur-reducing hyperthermophiles are associated to the first forms of life on Earth, the supply of sulfur in early times is supposed to have been more limited than now. However, due to geochemical evidence, it was proposed that iron could has been the first external electron acceptor in microbial metabolism [50]. **Table 2** shows a low concentration of sulfur in the ocean of Enceladus, and for this reason, the hydrothermal activity inferred by the molecular hydrogen detected from the plumes suggests that the iron could play a similar role in the geochemical reactions in the ocean of Enceladus.

Life not only came from hydrothermal vents but also, it could have risen on freshwater accumulations from geysers, precipitations, and hot spots, which could have linked to hydration-dehydration cycles. In hydrothermal vents, the thermal gradient allows for the concentration of solutes in the vents through the polymerization of minerals and sources of chemical energy like serpentinization. In the second system, the extreme concentration of chemical species took place due to the wetting-drying cycles, and the energy derived from evaporation provided the conditions of polymerization [51].

Enceladus looks like a potentially habitable world due to the similar current concentration of some key species present in the ocean to the ones that were present in the seawater of the oceans on Earth. There were detected traces of organic elements that could come from the water-rock interaction which can be also filled by minerals like iron, sodium, potassium, and calcium. There have been also detected the presence of biological consumable energy that on Earth, this energy is supplied by photosynthetic organisms like chemoautotrophs from a methanogenesis activity.

The environmental condition into the ocean of Enceladus could be in accord with life due to similarities with the oceans on Earth (pressures from 0.5 to 600 bar, which can be also found in some terrestrial environments [52], temperatures of 0–90°C, salinity calculated from the plumes between 0.5 and 2%. These values are lower than the ones on Earth which salinity is 3.5%). According to Porco et al. [53], the concentration of biological compounds could be potentially higher in the plume than in the seawater if the bubble scrubbing were allowed. These structures rise through the fluid while the organic material is attached to the water-gas interface until the eruption of the bubble through the jets. The addition of these organic compounds depends on their solubility and the surface activity. Surfactants like amphiphilic molecules would be instantly attached to the interface as they are able to reduce the surface tension, then the hydrophobic compounds would also be quickly attached.

Measurements in situ will be necessary to probe the feasibility of the ocean of Enceladus to harbor life. The information taken from the plumes by the Cassini mission provided data about the composition of the material expelled by the jets. The possibility to analyze samples from the plumes could bring a better understanding in how to make a characterization of the seawater and also, distinguish if there are residual elements that come from the interaction between living organisms and the environment.

5. Conclusions

The Earth had been through three different periods of time totally covered by ice, while maintaining a liquid ocean beneath the crust. The first Snowball event took place around 2.5 billion years ago, during the Paleoproterozoic age, and it was closely related to the Great Oxidation Event. The second and third Snowball events came about the Cryogenian era during the Neoproterozoic age, from 720 to 630 million years ago. Those Earth stages could be similar to the current physical and chemical conditions on Enceladus. The composition of the ocean on Enceladus is theorized through geochemical models, using the data taken by the Cassini mission. The concentration of species present in the material expelled from the plumes has been also calculated, allowing for the estimation of the pH of the ocean.

The pH is more basic on Enceladus than it was on Earth. The ratio of the carbonate equilibrium HCO_3^-/CO_3^{2-} was lower on Earth than on Enceladus. The most abundant ionic species are Na^+ and Cl^- in both oceans. The release of molecular hydrogen by serpentinization of the seafloor is also possible in both oceans. Sulphate species SO_4^{2-} appear to be scarcer on Enceladus but the role of electron acceptor could be taken by other elements like Iron, as it happened on Earth. The possible hydrothermal activity on Enceladus could be considered as a hint to biological activity, if it is compared with the hot spots from the deep oceans on Earth, where life arose.

The data calculated and compared in this research show a slightly similarity between the ocean on Enceladus and the oceans on Earth during the Snowball events, but it will be necessary to analyze some samples taken from the material expelled by the plumes. Previous research emphasized that the traces of organic material detected on Enceladus could come from biotic sources due to the few amino acids detected, that are known to be essential for the presence of life. Methane detected could also have a biotic origin, since there is a methanogenic bacterium called *Methanothermococcus okinawensis* which should be capable of thriving under the physical and chemical conditions of Enceladus. These organisms were found in a deep-sea hydrothermal vent, and they are able to survive in an environment with high temperatures and high pressures, up to 50 bar. The production of molecular hydrogen by serpentinization allows for them to survive in these extreme conditions. It is possible that these lifeforms can spread inside the hydrothermal activity that is present on Enceladus [54].

To probe the presence of biological activity on Enceladus and to infer the possible evolutionary primitive stage of its ocean, it is necessary to consider some bioindicators, such as the isotope carbon rates in organic and inorganic molecules, the ratio of simple hydrocarbons and amino acids in function of more complex molecules, and how the amino acids detected from the plumes could evolve. This research shows that the inorganic carbonates species are higher than the organic ones and the presence of sulphates are low, yet similar to the ones present in the oceans on Earth during the glaciation stages. Answering the question about the evolutionary stage of the ocean, these results allow us to speculate that, instead of having some keys species that could change the global conditions of Enceladus through time, it will be essential a global geological event that allows for the release of these species from the ocean to the surface, leading to an increase in the mass flow of species in the atmosphere and, therefore, an enrichment of it over time.

Furthermore, because of the presence of methane and some aminoacids, it could be possible to infer that, in the future, those molecules could evolve to more complex ones and ignite the chains of life. If more glaciations on Enceladus would happen in the future, it will allow the oxygenation of the atmosphere and the releasing of carbon dioxide into the atmosphere, leading to a change of the global conditions of Enceladus. It would be also important to analyze samples taken from the plumes, to have a better understanding of the seafloor conditions and to figure out which kind of extreme lifeforms could thrive on Enceladus.

Acknowledgements

This work was possible thanks to the Technology-Based Company RS3 Remote Sensing SL.

Conflict of interest

The authors declare no conflict of interest.

Author details

Katherine Villavicencio Valero[1,2]*, Emilio Ramírez Juidías[3] and Aina Àvila Bosch[4]

1 International Research School of Planetary Sciences, Università d'Annunzio, Pescara, Italy

2 Laboratorie de Géologie de Lyon Terre, Planètes, Environnement, Université Claude Bernard Lyon 1, Lyon, France

3 Graphic Engineering Department, University of Seville, ETSI, Seville, Spain

4 La Farga YourCopperSolutions S.A., Barcelona, Spain

*Address all correspondence to: katherine.villavicencio@univ-lyon1.fr

References

[1] Zeng Y, Jansen M. Ocean circulation on Enceladus with a high- versus Low-Salinity Ocean. The Planetary Science Journal. 2021;**2**:151. DOI: 10.3847/PSJ/ac1114

[2] Ray C, Glein C, Hunter J, Teolis B, Hoehler T, Huber J, et al. Oxidation processes diversify the metabolic menu on Enceladus. Icarus. 2020;**364**(4): 114248. DOI: 10.1016/j.icarus.2020.114248

[3] Hansen C, Esposito L, Colwell J, Hendrix A, Portyankina G, Stewart A, et al. The composition and structure of Enceladus' plume from the complete set of Cassini UVIS occultation observations. Icarus. 2020;**344**:113461. DOI: 10.1016/j.icarus.2019.113461

[4] Le Reun T, Hewitt D. Internally heated porous convection: An idealised model for Enceladus' hydrothermal activity. Journal of Geophysical Research: Planets. 2020;**125**: e2020JE00645. DOI: 10.1029/2020JE006451

[5] Rekier J, Trinh A, Triana S, Dehant V. Internal energy dissipation in Enceladus's Subsurface Ocean from tides and Libration and the role of inertial waves. Journal of Geophysical Research: Planets. 2019;**124**:2198-2212. DOI: 10.1029/2019JE005988

[6] Hsu H, Postberg F, Sekine Y, Shibuya T, Kempf S, Horányi M. Ongoing hydrothermal activities within Enceladus. Nature. 2015;**519**(7542): 207-210. DOI: 10.1038/nature14262

[7] Sekine Y, Shibuya T, Postberg F, Hsu H, Suzuki K, Masaki Y. High-temperature water–rock interactions and hydrothermal environments in the chondrite-like core of Enceladus. Nature Communications. 2015;**6**(1):1-8. DOI: 10.1038/ncomms9604

[8] Waite J, Lewis W, Magee B, Lunine J, McKinnon W, Glein C, et al. Liquid water on Enceladus from observations of ammonia and ^{40}Ar in the plume. Nature Letters. 2009;**460**:487-490. DOI: 10.1038/nature08153

[9] Khawaja N, Postberg F, Hillier J, Klenner F, Kempf S, Nolle L, et al. Low-mass nitrogen-, oxygen-bearing, and aromatic compounds in Enceladean ice grains. Monthly Notices of the Royal Astronomical Society. 2019;**489**: 5231-5243. DOI: 10.1093/mnras/stz2280

[10] Waite J, Combi M, Ip W, Cravens T, McNutt R, Kasprzak W, et al. Cassini ion and neutral mass spectrometer: Enceladus plume composition and structure. Science. 2006;**311**:1419-1422. DOI: 10.1126/science.1121290

[11] Seewald J. Detecting molecular hydrogen on Enceladus. Science. 2017; **356**:132-133. DOI: 10.1126/science.aan0444

[12] Postberg F, Clark R, Hansen C, Coates A, Dalle C, Scipioni F, et al. Plume and surface composition of Enceladus. In: Schenk P, Clark R, Howett C, Verbicer A, Waite J, editors. Handbook of Enceladus and the Icy Moons of Saturn. 1st ed. Tucson, USA: University of Arizona Press; 2018. pp. 129-162. DOI: 10.2458/azu_uapress_9780816537075-ch007

[13] Postberg F, Khawaja N, Abel B, Choblet G, Glein C, Gudipati M, et al. Macromolecular organic compounds from the depths of Enceladus. Nature Letters. 2018;**558**:564-568. DOI: 10.1038/s41586-018-0246-4

[14] Postberg F, Kempf S, Schmidt J, Brilliantov N, Beinsen A, Abel B, et al. Sodium salts in E-ring ice grains from an ocean below the surface of Enceladus. Nature Letters. 2009;**459**:1098-1101. DOI: 10.1038/nature08046

[15] Zolotov M. An oceanic composition on early and today's Enceladus. Geophysical Research Letters. 2007;**34**: L23203. DOI: 10.1029/2007GL031234

[16] Postberg F, Kempf S, Hillier J, Srama R, Green S, McBride N, et al. The E ring in the vicinity of Enceladus: II. Probing the moon's interior – The composition of E-ring particles. Icarus. 2008;**193**(2):438-454. DOI: 10.1016/j.icarus.2007.09.001

[17] Spencer J, Pearl C, Segura M, Flasar M, Mamoutkine A, Romani P, et al. Cassini encounters Enceladus: Background and the discovery of a south polar hot spot. Science. 2006;**311**: 1401-1405. DOI: 10.1126/science.1121661

[18] Choblet G, Tobie G, Sotin C, Běhounková M, Čadek O, Postberg F, et al. Powering prolonged hydrothermal activity inside Enceladus. Nature Letters. 2017;**1**:841-847. DOI: 10.1038/s41550-017-0289-8

[19] Choblet G, Tobie G, Sotin C, Kalousová K, Grasset O. Heat transport in the high-pressure ice mantle of large icy moons. Icarus. 2016;**285**:252-262. DOI: 10.1016/j.icarus.2016.12.002

[20] Woods P. Deep implications for H_2. Science. 2017;**356**:155-159. DOI: 10.1038/s41550-017-0136

[21] Waite J, Glein C, Perryman R, Teolis B, Magee B, Miller G, et al. Cassini finds molecular hydrogen in the Enceladus plume: Evidence for hydrothermal processes. Science. 2017; **356**:155-159. DOI: 10.1126/science.aai8703

[22] Glein C, Baross J, Waite J. The pH of Enceladus' ocean. Geochimica et Cosmochimica Acta. 2014;**162**:1-64. DOI: 10.1016/j.gca.2015.04.017

[23] Glein C, Waite J. The carbonate geochemistry of Enceladus' ocean. Geophysical Research Letters. 2020;**47**: e2019GL085885. DOI: 10.1029/2019GL085885

[24] Glein C, Postberg F, Vance S. The geochemistry of Enceladus: Composition and controls. In: Schenk P, Clark R, Howett C, Verbicer A, Waite J, editors. Handbook of Enceladus and the Icy Moons of Saturn. 1st ed. Tucson, USA: University of Arizona Press; 2018. pp. 39-56. DOI: 10.2458/azu_uapress_9780816537075-ch007

[25] Zandanel A, Truchea L, Hellmann R, Myagkiy A, Choblet G, Tobie G. Short lifespans of serpentinization in the rocky core of Enceladus: Implications for hydrogen production. Icarus. 2021;**364**: 114461. DOI: 10.1016/j.icarus.2021.114461

[26] Bouquet A, Mousis O, Waite J, Picaud S. Possible evidence for a methane source in Enceladus' ocean. GRJ Letters. 2015;**42**:1334-1339. DOI: 10.1002/2014GL063013

[27] Drabek E, Greaves J, Fraser H, Clements D, Alconcel L. Ground-based detection of a cloud of methanol from Enceladus: When is a biomarker not a biomarker? International Journal of Astrobiology. 2019;**18**:25-32. DOI: 10.1017/S1473550417000428

[28] Cable M, Porco C, Glein C, German C, MacKenzie S, Neveu M, et al. The science case for a return to Enceladus. The Planetary Science

Journal. 2021;**2**:132. DOI: 10.3847/PSJ/ABFB7A

[29] Anbar D, Duan Y, Lyons T, Arnold G, Kendall B, Creaser R, et al. A whiff of oxygen before the great oxidation event? Science. 2007;**317**: 1903-1906. DOI: 10.1126/science. 1140325

[30] Kopp R, Kirschvink J, Hilburn I, Nash C. The Paleoproterozoic snowball earth: A climate disaster triggered by the evolution of oxygenic photosynthesis. PNAS. 2005;**102**:11131-11136. DOI: 10.1073/pnas.0504878102

[31] Holland H. The oxygenation of the atmosphere and oceans. Philosophical Transactions of the Royal Society B: Biological Sciences. 2006;**361**: 903-915. DOI: 10.1098/rstb.2006. 1838

[32] Brown R, Clark R, Buratti B, Cruikshank D, Barnes J, Mastrapa R, et al. Composition and physical properties of Enceladus' surface. Science. 2004;**311**:1425-1428. DOI: 10.1126/ science.1121031

[33] Hoffman P, Schrag D. The snowball earth hypothesis: Testing the limits of global change. Terra Nova. 2002;**14**: 129-155. DOI: 10.1046/ j.1365-3121.2002.00408.x

[34] Paradise A, Menou K, Valencia D, Lee C. Habitable snowballs: Temperate land conditions, liquid water, and implications for CO_2 weathering. Journal of Geophysical Research: Planets. 2019;**8**:2087-2100. DOI: 10.1029/2019JE 005917

[35] Martin W, Baross J, Kelley D, Russell M. Hydrothermal vents and the origin of life. Nature. 2008;**6**:805-814. DOI: 10.1038/nrmicro1991

[36] Prieur D, Erauso G, Jeanthon C. Hyperthermophilic life at deep-sea hydrothermal vents. Planetary and Space Science. 1995;**42**:115-122. DOI: 10.1016/ 0032-0633(94)00143-F

[37] Sojo V, Herschy B, Whicher A, Camprubí E, Lane N. The origin of life in alkaline hydrothermal vents. Astrobiology. 2016;**16**:181-197. DOI: 10.1089/ast.2015.1406

[38] Couston L, Siegert M. Dynamic flows create potentially habitable conditions in Antarctic subglacial lakes. Science. 2021;**7**:eabc3972. DOI: 10.1126/ sciadv.abc3972

[39] Thomas P, Tajeddine R, Tiscareno M, Burns J, Joseph J, Loredo T, et al. Enceladus's measured physical libration requires a global subsurface ocean. Icarus. 2015;**264**:37-47. DOI: 10.1016/j.icarus.2015.08.037

[40] Kang W, Mittal T, Bire S, Campin J, Marshall J. How does salinity shape ocean circulation and ice geometry on Enceladus and other icy satellites? RS. 2021;**arXiv:2104.07008v3 [astro-ph. EP]**:1-64 DOI: 10.21203/rs.3.rs-143806/v1

[41] Ingersoll A, Nakajima M. Controlled boiling on Enceladus. 2. Model of the liquid-filled cracks. Icarus. 2016;**272**: 319-326. DOI: 10.1016/j.icarus.2015. 12.040

[42] Hansen C, Esposito L, Stewart A, Colwell J, Hendrix A, Pryor W, et al. Enceladus' water vapor plume. Science. 2006;**311**:1422-1425. DOI: 10.1126/ science.1121254

[43] Showalter M, Cuzzi J, Larson S. Structure and particle properties of Saturn's E ring. Icarus. 1991;**94**:451-473. DOI: 10.1016/0019-1035(91)90241-K

[44] Juhász A, Horányi M. Seasonal variations in Saturn's E-ring. Geophysical Research Letters. 2004;**31**: 19703. DOI: 10.1029/2004GL020999

[45] Southworth B, Kempf S, Schmidt J. Modeling Europa's dust plumes. Geophysical Research Letters. 2015;**42**: 541-548. DOI: 10.1002/2015GL066502

[46] Ramírez-Juidías E, Pozo-Morales L, Galán-Ortiz L. Procedure for obtaining a remote sensed image from a photograph. Patent n° ES2537783B2 (2015-09-29 publication of the patent concession). International Patent n° WO201419897 4A1. Universidad de Sevilla. 2015

[47] Millero F, Bruland K, Lohan M, Nightingale P, Liss P, de la Rocha C, et al. The oceans and marine geochemistry. In: Elderfield H, editor. Treatise on Geochemistry. Vol. 6. Oxford, UK: Elsevier; 2003

[48] Santosh M, Arai T, Maruyama S. Hadean earth and primordial continents: The cradle of prebiotic life. Geoscience Frontiers. 2017;**8**:309-327. DOI: 10.1016/j.gsf.2016.07.005

[49] Gibard C, Gorrell I, Jiménez E, Kee T, Pasek M, Krishnamurthy R. Geochemical sources and availability of Amidophosphates on the early earth. Angewandte Chemie. 2019;**131**: 8235-8239. DOI: 10.1002/ange.201903808

[50] Vargas M, Kashefi K, Blunt-harris E, Lovley D. Microbiological evidence for Fe(III) reduction on early earth. Nature. 1998;**395**:65-67. DOI: 10.1038/25720

[51] Deamer D, Damer B. Can life begin on Enceladus? A perspective from hydrothermal chemistry. Astrobiology. 2017;**17**:834-839. DOI: 10.1089/ast.2016.1610

[52] McKay C, Anbar A, Porco C, Tsou P. Follow the plume: The habitability of Enceladus. Astrobiology. 2014;**14**: 352-355. DOI: 10.1089/ast.2014.1158

[53] Porco C, Dones L, Mitchell C. Could it Be snowing microbes on Enceladus? Assessing conditions in its plume and implications for future missions. Astrobiology. 2017;**17**:876-901. DOI: 10.1089/ast.2017.1665

[54] Taubner R, Pappenreiter P, Zwicker J, Smrzka D, Pruckner C, Kolar P, et al. Biological methane production under putative Enceladus-like conditions. Nature. 2018;**9**:1-11. DOI: 10.1038/s41467-018-02876-y